Emerging trends in

CARBON EMISSION REDUCTION

JAGDISH KRISHANLAL ARORA

EMERGING TRENDS IN CARBON EMISSION REDUCTION
By
JAGDISH KRISHANLAL ARORA
techbagg@outlook.com

Also by Jagdish Krishanlal Arora

Table of Contents

Solar Energy

P erovskite Solar Cells: Perovskite solar cells have gained attention due to their potential to achieve higher efficiency and lower manufacturing costs compared to traditional silicon-based solar cells. Researchers have been working on improving their stability, scalability, and longevity.

Perovskite solar cells have emerged as promising alternatives to traditional silicon-based solar cells due to their potential for higher efficiency and lower manufacturing costs. Perovskite materials are crystalline structures that offer unique properties, including high light absorption efficiency and the ability to be processed into thin, lightweight, and flexible solar cells. Here are some key points regarding perovskite solar cells:

Advantages of Perovskite Solar Cells:

Higher Efficiency: Perovskite solar cells have demonstrated the potential to achieve higher power conversion efficiencies comparable to or exceeding those of silicon-based solar cells.

Lower Manufacturing Costs: Perovskite materials can be produced using simpler and more cost-effective fabrication techniques, including solution-based processes like inkjet printing or roll-to-roll manufacturing, reducing production expenses.

Versatility and Flexibility: Perovskite solar cells can be fabricated as thin films on various substrates, enabling their use in applications that require flexibility, such as integration into building materials or portable electronics.

Research Focus Areas for Improvement:

Stability and Longevity: Enhancing the stability and durability of perovskite materials under real-world operating conditions is a major focus. Addressing issues related to degradation from moisture, heat, and light exposure is crucial for commercial viability.

Scalability and Manufacturing Processes: Optimizing large-scale production methods and improving reproducibility in manufacturing are essential for scaling up perovskite solar cell production for commercial use.

Lead Toxicity and Material Composition: Many perovskite formulations contain lead, raising concerns about environmental impact and toxicity. Research aims to develop lead-free or less toxic alternatives while maintaining performance.

Recent Developments and Challenges:

Tackling Stability Issues: Researchers have made progress in stabilizing perovskite materials by introducing additives, encapsulation techniques, and developing novel material compositions to improve longevity.

Efficiency Improvements: Continuous efforts focus on boosting efficiency through tandem solar cell configurations, combining perovskite with other materials like silicon or improving light-trapping mechanisms.

Commercialization Hurdles: Commercializing perovskite solar cells faces challenges related to ensuring consistent performance, meeting industry standards, and establishing large-scale manufacturing processes.

Future Outlook:

The rapid advancements in perovskite solar cell technology show promise for revolutionizing the solar energy industry.

Continued research and development efforts aimed at improving stability, scalability, and environmental impact are key to realizing the full potential of perovskite solar cells and integrating them into mainstream renewable energy solutions. If the challenges regarding stability, scalability, and toxicity can be effectively addressed, perovskite solar cells could play a significant role in the global transition toward sustainable energy.

Bifacial Solar Panels: These panels can capture sunlight from both sides, increasing energy generation efficiency. Innovations in installation and tracking systems have improved their performance.

bifacial solar panels have garnered attention in the solar energy industry due to their ability to capture sunlight from both the front and rear sides, thereby enhancing energy generation efficiency compared to traditional single-sided solar panels. Here are some key points about bifacial solar panels and recent advancements:

Advantages of Bifacial Solar Panels:

Increased Energy Yield: Bifacial panels can generate additional electricity by capturing sunlight reflected off surfaces such as the ground, rooftops, or snow, enhancing overall energy output.

Versatility in Installation: These panels are adaptable to various mounting configurations, such as ground-mounted arrays, building-integrated installations, or canopies, maximizing their exposure to sunlight from both sides.

Improved Efficiency: Innovations in cell design, materials, and manufacturing processes have contributed to higher efficiency rates, further boosting the overall performance of bifacial solar panels.

Innovations in Installation and Tracking Systems:

Enhanced Mounting Systems: Advanced mounting structures, such as elevated or tilted setups, allow better airflow and minimize shading, optimizing the panel's ability to capture reflected light.

Innovations in installation and tracking systems for solar panels, particularly with advanced mounting structures, play a crucial role in maximizing the efficiency and performance of solar arrays, especially for bifacial panels. Here's more detail on these enhanced mounting systems:

Elevated Mounting Structures:

Increased Sun Exposure: Elevating solar panels off the ground, either through pole mounts or ground-based racks, reduces shading caused by nearby obstacles, vegetation, or snow accumulation. This allows for improved sunlight exposure to both sides of bifacial panels.

Improved Airflow: Elevating panels enhances airflow around the panels, reducing temperature build-up and improving heat dissipation. This helps maintain optimal operating temperatures, improving overall efficiency.

Tilted Setups:

Optimal Angles: Tilting solar panels at specific angles, adjustable based on geographical location and seasonal sun path, maximizes direct sunlight absorption and enhances the capture of reflected light from various surfaces.

Reduction in Soiling and Self-Cleaning: Tilted setups can facilitate natural cleaning of panels from rainfall or self-cleaning mechanisms, reducing dust or debris accumulation that might hinder sunlight absorption.

Benefits of Advanced Mounting Systems:

Increased Energy Output: Optimized mounting structures significantly improve the amount of light captured by solar panels, thereby increasing energy generation and overall efficiency.

Better Performance in Difficult Conditions: Elevated or tilted setups are advantageous in environments with uneven terrain, suboptimal ground conditions, or locations prone to shading, allowing for consistent and reliable energy production.

Technological Advances:

Smart Tracking Systems: Incorporating automated tracking systems that adjust the orientation of panels throughout the day to follow the sun's path optimizes energy capture, further enhancing overall efficiency.

Software-Assisted Designs: Advanced software and modelling tools aid in determining the best angles and configurations for mounting structures based on specific geographical and environmental conditions, maximizing energy yield.

Implementation and Practical Applications:

Large-Scale Solar Farms: Advanced mounting systems are widely utilized in utility-scale solar installations, where maximizing energy production is critical for economic viability.

Commercial and Residential Installations: These systems are increasingly applied in commercial buildings or residential properties, offering improved energy generation even in limited space or partially shaded areas.

Future Trends:

Continued advancements in mounting systems will likely focus on developing more adaptive, durable, and cost-effective solutions tailored to specific environmental conditions, aiming to further optimize solar panel performance and efficiency. As solar technology evolves, innovations in installation techniques will continue to play a pivotal role in enhancing the overall effectiveness of solar energy systems.

Tracking Technology: Utilizing solar tracking systems that follow the sun's path throughout the day can significantly increase energy production by ensuring the panels are always at an optimal angle to absorb sunlight.

solar tracking systems are pivotal in maximizing the efficiency and energy output of solar panels by adjusting their orientation to follow the sun's path across the sky throughout the day. Here's an overview of tracking technology and its benefits:

Types of Solar Tracking Systems:

Single-Axis Tracking: These systems adjust panels along one axis (either horizontal or vertical) to follow the sun's east-west path during the day, optimizing sunlight exposure.

Dual-Axis Tracking: Dual-axis systems adjust panels along both horizontal and vertical axes, allowing for more precise tracking of the sun's movement, optimizing for both azimuth and elevation angles.

Benefits of Solar Tracking Systems:

Increased Energy Yield: Solar trackers can enhance energy production by up to 20-25% compared to fixed-tilt systems by maintaining panels at an optimal angle to maximize sunlight absorption throughout the day.

Improved Efficiency: By continuously adjusting panel angles to minimize shading and maintain perpendicular alignment with sunlight, trackers ensure panels operate closer to their maximum efficiency.

How Solar Tracking Works:

Sensors and Motors: Solar trackers use sensors and motors to detect the position of the sun and adjust the orientation of panels accordingly, ensuring they face the sun directly for optimal exposure.

Control Systems: Automated control systems or algorithms calculate the sun's position based on time, date, and geographical location, controlling the movement of panels to track the sun's path accurately.

Applications and Use Cases:

Large-Scale Solar Farms: Solar tracking systems are commonly used in utility-scale solar installations and solar farms where maximizing energy output is crucial for cost-effectiveness.

Commercial and Industrial Installations: Some commercial and industrial installations also employ tracking systems to optimize energy production and increase overall system efficiency.

Challenges and Considerations:

Cost and Complexity: Solar tracking systems are generally more expensive and complex to install and maintain compared to fixed-tilt systems, which might impact the overall system cost.

Maintenance and Durability: Moving parts in tracking systems might require regular maintenance, and their exposure to environmental elements could affect durability and longevity.

Future Trends:

Continued advancements in solar tracking technology aim to address cost-efficiency, reliability, and durability concerns.

Research focuses on developing more cost-effective and robust tracking systems that balance increased energy production with manageable maintenance requirements. As solar tracking technology evolves, it will likely continue to play a crucial role in maximizing the efficiency and output of solar energy systems, especially in utility-scale installations where optimized energy generation is paramount.

Recent Developments:

Optimized Back sheet Design: Improvements in the rear-side encapsulation or back sheet material have increased durability, resistance to environmental factors, and minimized degradation, enhancing panel longevity.

Advancements in the design and materials used for the back sheet, which is the rear-side encapsulation of solar panels, have contributed significantly to improving panel durability, resistance to environmental factors, and overall longevity. Here's a breakdown of optimized back sheet design and its benefits:

Importance of Back sheets:

Protection and Insulation: The back sheet serves as a protective layer, insulating the solar cells from environmental factors such as moisture, humidity, temperature fluctuations, and UV radiation.

Electrical Insulation: It provides electrical insulation, preventing electrical leakage and ensuring safe operation of the solar panel.

Advancements in Back sheet Design:

Enhanced Durability: Improvements in materials and construction techniques have led to back sheet designs that offer higher resistance to wear, tearing, and mechanical stress, contributing to longer panel lifespans.

Improved Weather Resistance: Advanced back sheet materials are formulated to withstand harsh weather conditions, UV exposure, and temperature variations, reducing the risk of degradation over time.

Benefits of Optimized Back sheet Design:

Extended Panel Lifespan: Durable back sheet materials enhance the longevity of solar panels, ensuring they can withstand environmental stresses and maintain performance over an extended period.

Reduced Degradation: Enhanced weather resistance minimizes the effects of moisture ingress, UV exposure, and thermal cycling, reducing the risk of back sheet degradation that could otherwise impact panel efficiency.

Material Innovations:

Tedlar-Polyester-Tedlar (TPT) and Tedlar-Polyvinyl Fluoride (PVF)-Based Back sheets: These materials provide excellent weather resistance, durability, and electrical insulation, commonly used in high-quality solar panels.

Fluoropolymer-Based Back sheets: Materials like fluoropolymer films offer improved UV stability, weather resistance, and thermal stability, contributing to enhanced panel durability.

Research and Development:

Focus on Long-Term Performance: Continued R&D efforts aim to develop back sheet materials that maintain their properties and protect solar panels over an extended period, ensuring consistent performance.

Environmental Considerations: There's a push for developing environmentally friendly back sheet materials with reduced environmental impact during production, usage, and end-of-life disposal or recycling.

Application and Impact:

Commercial and Residential Installations: Improved back sheet designs are incorporated into both commercial and residential solar panels, ensuring longer-lasting and more reliable systems.

Utility-Scale Solar Projects: Utility-scale solar installations benefit from optimized back sheet designs by ensuring sustained performance and reducing maintenance requirements.

Future Trends:

As the solar industry continues to evolve, further advancements in back sheet materials and designs will likely focus on enhancing durability, environmental sustainability, and cost-effectiveness. Innovations will aim to strike a balance between increased durability and maintaining or reducing overall panel costs, ensuring that solar panels remain a durable and reliable renewable energy source.

Advanced Cell Technology: Ongoing research focuses on developing more efficient and cost-effective cell technologies, including heterojunction or passivated emitter rear contact (PERC) cells, to further enhance panel performance.

Ongoing research and development in solar cell technology aim to enhance efficiency, cost-effectiveness, and overall performance of solar panels. Innovations like heterojunction and passivated emitter rear contact (PERC) cells have shown significant promise in achieving these goals. Here's an overview of these advanced cell technologies:

Heterojunction Solar Cells:

Working Principle: Heterojunction cells use multiple layers of different semiconductor materials. They incorporate amorphous silicon and crystalline silicon, improving energy conversion efficiency by reducing energy losses due to recombination.

Efficiency and Performance: Heterojunction cells have demonstrated higher efficiency rates compared to traditional silicon-based cells, achieving better conversion of sunlight into electricity.

Passivated Emitter Rear Contact (PERC) Cells:

Enhanced Rear Surface Passivation: PERC cells incorporate a passivation layer at the rear surface of the cell, reducing electron recombination losses and improving overall efficiency.

Increased Light Absorption: PERC technology enables better light absorption, particularly in longer wavelengths, improving the cell's ability to convert sunlight into electricity.

Benefits of Advanced Cell Technologies:

Higher Efficiency: Both heterojunction and PERC cells have shown the potential to achieve higher conversion efficiencies compared to conventional solar cells, leading to increased energy production.

Improved Performance in Low-Light Conditions: These technologies exhibit better performance in low-light conditions, allowing for more consistent energy generation even during overcast or cloudy days.

Research Focus Areas:

Cost Reduction: Continuous research aims to streamline manufacturing processes and materials used in these advanced cell technologies to reduce production costs and make them more economically viable.

Durability and Longevity: Efforts are focused on enhancing the durability and reliability of these cells to ensure prolonged performance over their operational lifespan.

Commercial Adoption:

Utility-Scale Installations: These advanced cell technologies are increasingly adopted in utility-scale solar projects due to their higher efficiency, leading to improved energy yields and reduced installation costs per watt.

Residential and Commercial Applications: Heterojunction and PERC cells are also being integrated into residential and commercial solar systems, providing better energy generation in limited space.

Future Developments:

Research efforts continue to advance these cell technologies, aiming to push the efficiency boundaries further while optimizing their manufacturing processes for larger-scale adoption. Additionally, innovations may focus on integrating these advanced cells with emerging technologies, such as tandem or multi-junction solar cells, to achieve even higher efficiencies.

The continuous development and integration of these advanced cell technologies contribute significantly to the ongoing evolution and improvement of solar panels, making them more efficient, cost-effective, and reliable for widespread use in renewable energy generation.

Benefits in Various Applications

Commercial and Utility-Scale Projects: Bifacial solar panels are increasingly used in large-scale solar installations, providing enhanced energy output and better performance in locations with favourable conditions for reflected light.

The utilization of bifacial solar panels in both commercial and utility-scale projects has been on the rise due to their ability to enhance energy output and performance, especially in locations conducive to capturing reflected light. Here's an overview of their applications in such projects:

Advantages in Commercial and Utility-Scale Installations:

Increased Energy Generation: Bifacial solar panels can capture sunlight from both sides, including direct sunlight on the front and reflected light from the rear, significantly boosting overall energy production.

Optimized for Large Installations: In utility-scale solar farms or commercial projects with extensive installations, bifacial panels can capitalize on expansive areas and maximize energy yield from varied surfaces.

Benefits in Large-Scale Projects

Enhanced Efficiency in Open Spaces: Bifacial panels perform well in open spaces with reflective surfaces like white gravel, snow, or desert sands, where they can capture additional sunlight reflected from the ground.

Reduced LCOE (Levelized Cost of Energy): The increased energy output per panel helps lower the levelized cost of energy, making bifacial panels an attractive option for cost-efficient large-scale projects.

Site Suitability and Performance Factors

Sunlight Reflection: Locations with surfaces that reflect sunlight, such as snow-covered areas, light-coloured ground coverings, or highly reflective materials, provide optimal conditions for bifacial panels.

Orientation and Tilt: Mounting and tilting configurations are crucial; these panels perform best in locations where the ground reflects light onto the rear side of the panels, requiring proper orientation and elevation angles.

Technology Integration

Advanced Mounting Structures: Elevated or tilted setups, coupled with smart tracking systems, further optimize bifacial panel performance by maximizing exposure to reflected light and adjusting panel angles for optimal sunlight absorption.

Data-Driven Design and Planning: Utilizing data analytics and modelling tools assists in determining the most suitable locations and configurations for installing bifacial panels to maximize energy output.

Commercial Adoption and Future Trends:

Growing Market Share: The adoption of bifacial solar panels in utility-scale projects is increasing, driven by their proven performance and cost-effectiveness, leading to a larger market share within the solar industry.

Continued Technological Advancements: Ongoing research focuses on improving the reliability, durability, and efficiency of bifacial panels, making them more competitive and adaptable for various environmental conditions.

Regulatory Support and Industry Standards:

Recognition and Standards: Industry recognition and support through standards and regulations that acknowledge the performance and potential benefits of bifacial solar panels further encourage their use in commercial and utility-scale installations.

The increasing deployment of bifacial solar panels in large-scale projects reflects their ability to enhance energy generation and provide cost-effective solutions, especially in locations with favourable conditions for capturing reflected light. Continued advancements and improvements in technology are expected to further solidify their position as a viable and efficient option in the solar energy sector.

Residential and Architectural Integration: These panels offer design flexibility and can be integrated into building structures or rooftops, making them suitable for residential and architectural applications.

bifacial solar panels offer design flexibility and integration possibilities that make them well-suited for residential and architectural applications. Their unique characteristics allow for seamless incorporation into building structures or rooftops, offering both aesthetic appeal and functional energy generation. Here's an overview of their suitability and advantages in residential and architectural integration:

Design Flexibility and Aesthetic Appeal:

Versatile Integration: Bifacial panels can be integrated into various architectural designs, such as building facades, awnings, canopies, or as part of the roofing material, due to their thin and flexible construction.

Transparency Options: Some bifacial panels can be designed with varying levels of transparency, enabling the creation of semi-translucent installations that allow diffused light while generating electricity.

Rooftop Applications

Rooftop Installations: Bifacial panels integrated into rooftops harness both direct sunlight and reflected light, optimizing energy production while serving as a part of the building structure.

Aesthetic Considerations: Their sleek design and adaptability enable homeowners and architects to incorporate solar technology without compromising the overall aesthetic appeal of the building.

Architectural Integration:

Building-Integrated PV (BIPV): Bifacial panels can serve as an integral part of the building envelope, replacing conventional building materials like glass, siding, or roofing while generating electricity.

Customizable Solutions: Panels can be tailored in size, shape, or colour to complement specific architectural designs or aesthetics, enhancing the overall visual appeal of the structure.

Benefits for Residential Use

Space Efficiency: Bifacial panels optimize space utilization, making them suitable for homes with limited rooftop area by generating additional power from both sides.

Energy Cost Savings: Residential installations benefit from reduced energy costs by generating clean electricity and potentially feeding excess power back into the grid.

Future Trends and Adoption

Consumer Demand: Increasing interest in sustainable living and renewable energy solutions is driving consumer demand for solar panels that offer both functionality and aesthetic appeal.

Technological Advancements: Ongoing developments aim to improve the aesthetics, efficiency, and durability of bifacial panels, making them even more attractive for residential and architectural integration.

Regulatory Support and Incentives

Incentive Programs: Government incentives, tax credits, and rebates for residential solar installations encourage homeowners to invest in solar energy, further promoting the adoption of bifacial panels.

Building Regulations: Supportive policies and building regulations that promote the integration of solar technology into construction projects encourage architects and developers to consider bifacial panels as viable options.

Sustainability and Longevity

Green Building Certifications: Bifacial panels align with green building certifications, promoting sustainable construction practices and contributing to environmentally friendly structures.

Long-Term Investment: As technology improves and prices become more competitive, the long-term durability and energy efficiency of bifacial panels make them a sound investment for homeowners and building developers alike.

The integration of bifacial solar panels into residential and architectural designs presents a harmonious blend of functionality, aesthetics, and sustainability, offering a promising avenue for clean energy adoption while enhancing the visual appeal of buildings and homes. Continued advancements and wider adoption of these panels are expected to further propel their role in the architectural landscape and residential solar market.

Future Prospects

As the solar industry continues to evolve, advancements in materials, manufacturing techniques, and system optimization are expected to further improve the efficiency, reliability, and cost-effectiveness of bifacial solar panels. Their ability to generate more electricity from the same footprint makes them a promising technology, especially in locations with ample sunlight and favourable surface conditions for reflection. As research and development efforts progress, bifacial solar panels are poised to play a significant role in the expansion of renewable energy generation.

the evolution of bifacial solar panels represents a promising trajectory within the solar industry, poised to make a substantial impact on renewable energy generation. Advancements in materials, manufacturing techniques, and system optimization are key drivers in enhancing the efficiency, reliability, and cost-effectiveness of bifacial panels. Here's a deeper dive into the potential and ongoing developments:

Continued Improvement in Efficiency and Reliability:

Materials Innovation: Research focuses on developing novel materials with improved light absorption properties, better durability, and reduced degradation rates, enhancing overall panel performance.

Optimized Manufacturing Techniques: Streamlining production processes, such as improving cell interconnection methods and reducing material waste, aims to increase production efficiency and reduce costs.

System Optimization for Enhanced Performance:

Advanced Mounting Configurations: Further refinement of mounting structures, including tilted or elevated setups, coupled with improved tracking systems, maximizes the capture of reflected light, enhancing energy output.

Smart Analytics and Design Tools: Data-driven approaches and modelling tools help in optimizing installation layouts and configurations to maximize energy generation in diverse environmental conditions.

Tailored Solutions for Diverse Environments:

Geographical Adaptability: Ongoing R&D aims to develop bifacial panels tailored for specific environmental conditions, ensuring optimal performance in various locations, from desert regions to snowy climates.

Climate Resilience: Panels designed to withstand extreme weather conditions, including high temperatures, humidity, snow, and hail, are crucial for ensuring long-term performance and reliability.

Market Expansion and Economic Viability:

Cost Reduction: Continued advancements target cost reductions through economies of scale, improved manufacturing efficiency, and the adoption of more cost-effective materials and technologies.

Market Penetration: Increasing competitiveness and favourable cost-to-efficiency ratios are expected to drive greater market penetration, expanding the adoption of bifacial panels across diverse applications.

Technological Convergence and Integration:

Tandem Cell Configurations: Research explores integrating bifacial panels with other emerging technologies, such as tandem cell configurations, to achieve higher efficiencies and optimize energy generation.

Energy Storage Integration: Bifacial panels integrated with energy storage solutions aim to provide round-the-clock renewable energy access, further enhancing their appeal and utility.

Policy Support and Industry Collaboration:

Regulatory Backing: Supportive policies, incentives, and regulatory frameworks that recognize the efficiency and potential of bifacial panels encourage their deployment in various markets.

Collaborative Initiatives: Industry partnerships and collaborations foster innovation, knowledge-sharing, and standardization, accelerating the pace of technological advancements.

Environmental Impact and Sustainability:

Lifecycle Assessment: Continued efforts aim to minimize environmental impact through sustainable manufacturing practices and recycling programs, ensuring a positive lifecycle footprint.

Green Certifications: Bifacial panels align with sustainability certifications and green building standards, driving their adoption in eco-friendly construction projects.

Future Outlook:

As research and development progress, bifacial solar panels are expected to continue playing a pivotal role in the global expansion of renewable energy. Their ability to generate more electricity from the same footprint, coupled with ongoing improvements in efficiency, reliability, and cost-effectiveness, positions them as a key technology for the sustainable transformation of the energy sector. The convergence of technological advancements, supportive policies, and market demand further solidifies their potential as a significant contributor to the clean energy landscape.

Solar Energy Storage

Solar Energy Storage: Advancements in battery technologies, such as lithium-ion batteries and flow batteries, aim to store excess solar energy for use during low sunlight periods or at night, improving overall solar energy utilization.

Advancements in energy storage technologies, particularly in batteries like lithium-ion batteries and flow batteries, are crucial for maximizing the utilization of solar energy by storing excess power generated during periods of sunlight for use during low-light conditions or at night. Here's an overview of how these battery technologies contribute to solar energy storage:

Lithium-Ion Batteries:

Efficient Energy Storage: Lithium-ion batteries are widely used for storing solar energy due to their high energy density, efficiency, and relatively low maintenance requirements.

Scalability and Adaptability: These batteries are adaptable for various system sizes, from small residential installations to utility-scale projects, making them versatile for solar energy storage.

Fast Response Time: Lithium-ion batteries have quick response times, allowing rapid discharge when stored energy is required, making them suitable for addressing fluctuations in solar energy availability.

Flow Batteries:

Longer Duration Storage: Flow batteries, such as vanadium redox flow batteries, excel in longer-duration energy storage, offering scalability and the ability to store large amounts of energy.

Decoupled Power and Capacity: Flow batteries can separate power and capacity, providing flexibility in adjusting the storage size and power output independently.

Enhanced Durability: Flow batteries have the potential for longer lifespans compared to some other battery technologies, making them suitable for applications requiring extended cycle life.

Benefits of Solar Energy Storage:

Increased Self-Sufficiency: Energy storage allows solar-powered homes or facilities to rely less on the grid by storing excess solar energy for use during times of low sunlight or at night.

Grid Stabilization: Storage systems help in grid stabilization by mitigating fluctuations caused by intermittent renewable energy sources, ensuring a more reliable energy supply.

Technological Advancements:

Improved Efficiency and Performance: Ongoing R&D focuses on enhancing battery efficiency, extending cycle life, reducing degradation rates, and increasing energy density for better overall performance.

Cost Reduction: Efforts to reduce the cost of materials, manufacturing, and scaling up production contribute to making energy storage more economically viable.

Integration with Solar Installations:

Residential and Commercial Integration: Energy storage systems are increasingly integrated into residential and commercial solar setups, offering users greater control over their energy consumption and reducing electricity bills.

Utility-Scale Applications: Large-scale solar farms and utility-scale projects integrate energy storage systems to provide more stable and dispatchable renewable power to the grid.

Future Outlook:

Continued advancements in battery technologies, coupled with ongoing research and innovation, are expected to drive further improvements in energy storage efficiency, cost-effectiveness, and reliability. As the demand for renewable energy grows, the integration of advanced energy storage solutions with solar power will continue to play a pivotal role in fostering a more sustainable and resilient energy ecosystem.

Wind Energy

Turbine Efficiency: Ongoing research focuses on enhancing the efficiency of wind turbines through improved blade designs, materials, and aerodynamics. Larger turbines are being developed to capture more wind energy at higher altitudes.

Since, ongoing research and development efforts in the field of wind energy are primarily focused on enhancing the efficiency of wind turbines. Advancements in blade designs, materials, aerodynamics, and the scale of turbines are crucial aspects contributing to increased efficiency. Here's an overview of these ongoing improvements:

Improved Blade Designs:

Advanced Aerodynamics: Research aims to optimize blade shapes and profiles to improve aerodynamic efficiency, reducing drag and increasing lift to capture more wind energy.

Therefore, enhancing wind turbine blade designs through advanced aerodynamics plays a pivotal role in improving their efficiency, increasing energy capture, and reducing overall costs. Here's a deeper dive into how research focuses on optimizing blade shapes and profiles to achieve improved aerodynamic efficiency:

AERODYNAMIC OPTIMIZATION:

Streamlined Shapes: Researchers are working on blade profiles that minimize air resistance (drag) by employing sleek and aerodynamic shapes, reducing the force that opposes wind flow.

Airfoil Design Enhancement: Optimizing the cross-sectional shape (airfoil) of wind turbine blades to maximize lift, enabling them to efficiently harness more energy from the passing wind.

Turbulence Mitigation:

Noise Reduction: Improved blade designs aim to mitigate noise generated by wind turbines through aerodynamic modifications, reducing turbulence and resulting in quieter operation.

Vortex Generators: Implementing small aerodynamic structures, like vortex generators, on blades helps control airflow separation, enhancing lift and overall efficiency.

Computational Fluid Dynamics (CFD):

Simulation and Modelling: Computational tools simulate airflow around blades, aiding in the development of optimized shapes, profiles, and surface textures for improved aerodynamic performance.

Virtual Testing: CFD simulations allow for virtual testing of various blade designs, enabling faster iteration cycles and cost-effective refinement before physical prototyping.

Blade Twist and Angle of Attack:

Optimized Twist Distribution: Research focuses on adjusting the blade twist along its length to maintain optimal angles of attack across different wind speeds, enhancing efficiency.

Angle Adjustment Mechanisms: Incorporating adaptive mechanisms to alter blade angles in real-time based on wind conditions allows for continuous optimization of energy capture.

Wind Tunnel Testing:

Physical Prototyping: Wind tunnel experiments and physical testing validate and refine computational models, ensuring that the proposed designs exhibit the predicted aerodynamic performance.

Validation of Designs: Wind tunnel tests provide empirical data on airflow characteristics, helping validate the efficacy of new blade designs in controlled environments.

Industry Adoption and Impact:

Improved Efficiency: Enhanced aerodynamics translate to increased turbine efficiency, resulting in higher energy production and lower operational costs over the turbine's lifespan.

Commercial Applications: Advancements in blade design and aerodynamics are gradually implemented in newer wind turbine models, contributing to the overall advancement of wind energy technology.

Future Prospects:

Ongoing research in aerodynamics aims to further optimize blade designs, pushing the boundaries of efficiency and performance. As technology evolves, innovative aerodynamic enhancements will continue to be crucial in maximizing the energy capture potential of wind turbines, making wind power a more reliable and competitive source of renewable energy.

Smart Blade Technology: Incorporation of sensors, microcontrollers, and adaptive materials in blade design allows for real-time adjustments, optimizing performance under varying wind conditions.

Smart blade technology represents a significant advancement in wind turbine design by integrating sensors, microcontrollers, and adaptive materials to enable real-time adjustments, enhancing overall performance under varying wind conditions. Here's an in-depth look at how this technology works and its benefits:

Components of Smart Blade Technology:

Sensors and Monitoring Systems:

Strain gauges, accelerometers, and other sensors are embedded within the blade structure to monitor factors like blade deformation, loads, vibrations, and environmental conditions.

Microcontrollers and Control Systems:

Microprocessors receive data from sensors and interpret it, making real-time decisions to optimize blade performance or adjust operational parameters.

Adaptive Materials or Mechanisms:

Usage of materials with adaptive properties (e.g., shape memory alloys) or mechanisms that can change the blade shape or characteristics based on incoming data.

Functionality and Benefits:

Real-Time Performance Optimization:

Sensors collect data on wind speed, direction, turbulence, and loads, enabling microcontrollers to make immediate adjustments for optimal performance.

Load Mitigation:

Smart blades can adjust their orientation or stiffness dynamically to mitigate loads caused by turbulent wind conditions, enhancing structural integrity and longevity.

Efficiency Improvement:

Adaptive materials or mechanisms allow the blade to dynamically change its shape or angle, optimizing energy capture in varying wind speeds or directions.

Fault Detection and Maintenance:

Continuous monitoring helps identify potential faults or issues early, allowing for predictive maintenance and minimizing downtime due to blade-related problems.

Adaptive Mechanisms:

Active Flap Systems:

Adjustable sections on the blade's trailing edge can change the angle or orientation, optimizing aerodynamic performance in real-time.

Twist Control Mechanisms:

Mechanisms that adjust the twist along the blade length, optimizing aerodynamic efficiency and reducing loads in changing wind conditions.

Research and Development:

Material Innovation:

Development of smart materials with properties that respond dynamically to environmental changes, enabling blades to adapt to varying conditions.

Algorithmic Optimization:

Refinement of control algorithms that interpret sensor data and make precise adjustments to optimize energy capture and turbine performance.

Industry Integration and Future Outlook:

Prototyping and Testing:

Wind turbine manufacturers are testing and refining smart blade technologies in prototype turbines to validate their effectiveness and reliability.

Potential for Future Deployment:

Continued advancements may see wider adoption of smart blade technology in commercial wind turbines, contributing to increased efficiency and reliability in wind energy generation.

Environmental Impact and Sustainability:

Improved Efficiency, Reduced Maintenance: Smart blade technology contributes to reducing maintenance needs and increasing turbine efficiency, making wind energy more sustainable and cost-effective.

Enhanced Reliability: By actively adjusting to changing conditions, smart blades can potentially extend turbine lifespan, reducing the environmental impact associated with replacing components.

Smart blade technology represents an exciting frontier in wind turbine innovation, offering the potential to significantly enhance efficiency, reliability, and sustainability in wind energy generation by enabling turbines to adapt and optimize their performance in real-time, ultimately contributing to a more efficient and reliable renewable energy source.

Advanced Materials:

Composite Materials: Use of lighter yet stronger composite materials like carbon fiber or fiberglass in blade construction improves durability and allows for longer, more aerodynamically efficient blades.

the incorporation of advanced composite materials, such as carbon fiber or fiberglass, in wind turbine blade construction has revolutionized the industry by offering lighter, stronger, and more durable blade designs. Here's an overview of how these composite materials enhance wind turbine blades:

Composite Materials Advantages:

Strength-to-Weight Ratio:

Carbon fiber and fiberglass offer exceptional strength while being significantly lighter than traditional materials like steel or aluminum, allowing for larger, more efficient blades.

Enhanced Durability:

These materials exhibit high resistance to fatigue, corrosion, and environmental degradation, ensuring longer blade lifespans and reduced maintenance needs.

Flexibility in Design:

Composite materials allow for more intricate and optimized blade shapes, enabling designers to create aerodynamically efficient profiles for better energy capture.

Carbon Fiber:

High-Strength Properties:

Carbon fiber composites possess exceptional strength properties, making them ideal for withstanding high loads and stresses experienced by wind turbine blades.

Low Weight and Stiffness:

The low weight-to-stiffness ratio of carbon fiber allows for lighter yet highly rigid blades, reducing material fatigue and enabling longer and more efficient blades.

Fiberglass:

Cost-Effectiveness:

Fiberglass composites offer a cost-effective alternative with good strength and durability, often used in smaller-scale wind turbines and various structural applications.

Versatility in Manufacturing:

Fiberglass materials can be moulded into various shapes and sizes, providing versatility in manufacturing and allowing for complex blade designs.

Impact on Blade Performance:

Increased Energy Capture:

Lighter and longer blades made from composite materials capture more wind energy, contributing to increased turbine efficiency and power output.

Reduced Structural Fatigue:

Composite materials' durability reduces structural fatigue caused by constant wind loads and stress, enhancing blade longevity and reliability.

Manufacturing and Challenges:

Manufacturing Processes:

Techniques like vacuum infusion or resin transfer moulding are used to fabricate composite blades, ensuring precise control over material distribution and quality.

Scaling and Cost Considerations:

While composite materials offer superior performance, challenges related to scaling production and initial costs need to be addressed for wider industry adoption.

Future Developments:

Continuous Research: Ongoing research focuses on improving composite materials and manufacturing techniques to reduce costs and increase scalability.

Sustainable Composites: Development of sustainable and recyclable composite materials for wind turbine blades to further reduce environmental impact and end-of-life disposal concerns.

Environmental Impact:

Energy Efficiency and Sustainability: Lightweight composite blades contribute to higher energy production efficiency, ultimately promoting the overall sustainability of wind energy.

Recycling Initiatives: Efforts are underway to develop recycling methods for composite materials to minimize waste and environmental impact at the end of a blade's lifespan.

The utilization of advanced composite materials like carbon fiber and fiberglass in wind turbine blades has revolutionized the industry by offering stronger, lighter, and more durable solutions. These materials contribute significantly to increased energy capture, improved efficiency, and enhanced sustainability in wind energy generation, paving the way for more efficient and reliable renewable energy sources.

Smart Coatings: Surface coatings with reduced friction or self-cleaning properties enhance airflow over the blades, reducing turbulence and improving efficiency.

smart coatings play a pivotal role in enhancing the performance and efficiency of wind turbine blades by applying specialized surface treatments that offer reduced friction or self-cleaning properties. These coatings contribute to optimizing airflow over the blades, reducing turbulence, and ultimately improving overall efficiency. Here's a closer look at how smart coatings benefit wind turbine blades:

Reduced Friction Coatings:

Improved Aerodynamics:

Coatings with reduced friction properties help streamline airflow over the blade surface, minimizing drag and enhancing aerodynamic efficiency.

Enhanced Energy Capture:

By reducing resistance, these coatings enable smoother airflow, allowing the blades to capture more wind energy and increase overall turbine output.

Self-Cleaning Properties:

Dirt and Debris Resistance:

Self-cleaning coatings repel dust, dirt, or environmental debris, preventing buildup on the blade surface that could disrupt airflow and efficiency.

Maintenance Reduction:

Coatings that repel contaminants reduce the need for manual cleaning and maintenance, leading to decreased downtime and operational costs.

Anti-Icing or De-Icing Coatings:

Prevention of Ice Buildup:

Special coatings can prevent or reduce ice formation on blade surfaces, mitigating the negative impact of icing on turbine efficiency and operation.

Enhanced Reliability:

Anti-icing coatings help maintain turbine functionality during cold weather, ensuring consistent energy generation even in icy conditions.

Nanotechnology and Surface Modification:

Nanostructured Coatings:

Utilization of nanotechnology allows for the creation of coatings with micro- or nanostructured surfaces that optimize aerodynamics and reduce adhesion of contaminants.

Surface Bonding Modifications:

Surface treatments alter the blade material's surface properties, reducing the likelihood of dirt, ice, or water adherence, leading to self-cleaning or de-icing capabilities.

Impact on Turbine Efficiency:

Optimized Airflow:

Improved blade surfaces through smart coatings reduce drag, turbulence, and resistance, enhancing the conversion of wind energy into electrical power.

Increased Energy Yield:

Enhanced aerodynamics and reduced surface contamination result in increased energy capture and higher turbine output over time.

Environmental Impact and Sustainability:

Reduced Environmental Contamination:

Coatings that prevent contaminants from adhering to the blades minimize the risk of environmental pollution from dust, dirt, or ice buildup.

Long-Term Performance:

Smart coatings contribute to longer-lasting blades by protecting against wear, corrosion, or damage, ultimately enhancing the sustainability of wind turbine operations.

Future Developments:

Research on Advanced Coatings:

Ongoing research aims to develop more durable, cost-effective, and environmentally friendly smart coatings tailored specifically for wind turbine applications.

Integration with Blade Materials:

Further advancements seek to integrate these coatings directly into blade manufacturing processes, ensuring their longevity and effectiveness.

Smart coatings with reduced friction, self-cleaning, or anti-icing properties offer significant advantages in improving the efficiency and performance of wind turbine blades. These coatings contribute to increased energy capture, reduced maintenance needs, and enhanced sustainability, ultimately advancing the viability and effectiveness of wind energy as a clean and renewable power source.

Aerodynamic Enhancements:

Leading Edge Modifications: Innovative designs for the leading edge of turbine blades aim to minimize drag and increase lift, optimizing energy capture from the wind.

Turbine Configuration: Research explores various turbine configurations, including rotor diameter, number of blades, and hub height, to maximize efficiency at different wind speeds and locations.

Larger Turbines and Higher Altitudes:

Scale and Size: Development of larger turbines with increased rotor diameters captures more wind energy, while taller towers elevate turbines to higher altitudes with stronger and more consistent wind speeds.

Floating Turbines: Advancements in floating offshore wind turbines enable deployment in deeper waters, accessing stronger and more consistent winds for increased energy production.

Benefits of Enhanced Efficiency:

Increased Energy Yield: Improved turbine efficiency leads to higher energy yields, enhancing overall electricity generation from wind power installations.

Cost Reduction: Higher efficiency turbines reduce the levelized cost of energy (LCOE), making wind energy more competitive compared to conventional energy sources.

Future Outlook:

Continued research and innovation in wind turbine technology are expected to drive further improvements in efficiency and cost-effectiveness. As larger turbines with advanced designs become more prevalent, they will contribute significantly to meeting renewable energy goals by harnessing more wind power efficiently. Additionally, advancements in offshore wind technology, including floating turbines and increased installation capacities, hold promise for expanding wind

41

energy generation to new geographical locations and further increasing overall efficiency and capacity factors.

Advancements in Efficiency and Cost-Effectiveness:

Innovative Designs: Continued research aims to refine blade shapes, materials, and aerodynamics, leading to higher efficiency in energy capture and reduced costs per unit of energy produced.

Smart Technologies: Integration of smart systems, such as advanced sensors, predictive maintenance, and control algorithms, will optimize turbine performance and reduce operational costs.

Scaling Up Production: Streamlining manufacturing processes and increasing economies of scale will help reduce the upfront costs of wind turbines, making them more cost-effective.

Larger Turbines and Advanced Designs:

Larger Rotor Diameters: Development of larger turbines with increased rotor sizes will harness more wind energy, improving capacity factors and overall energy production.

Taller Towers: Elevated turbines reaching higher altitudes will access stronger and more consistent winds, enhancing energy yield and operational efficiency.

Offshore Wind Advancements:

Floating Turbines: Advancements in floating offshore wind technology will enable the deployment of turbines in deeper waters, expanding the potential for offshore wind farms in new areas.

Increased Installation Capacities: Continued investments and technological advancements will lead to larger offshore wind installations with higher capacity factors, contributing significantly to energy production.

Grid Integration and Energy Storage:

Grid-Friendly Solutions: Technologies for better grid integration, such as energy storage systems and demand-side management, will ensure stability and reliable energy supply.

Hybrid Solutions: Integration of wind power with complementary technologies like solar, energy storage, or hybrid renewable systems will enhance reliability and energy availability.

Environmental Impact and Sustainability:

Reduced Footprint: Advanced designs and improved efficiency will lead to a reduced environmental footprint per unit of energy generated, making wind power more sustainable.

Lifecycle Assessments: Ongoing research will focus on the lifecycle analysis of wind turbines, ensuring their sustainability from manufacturing to decommissioning.

Global Expansion and Accessibility:

New Geographic Locations: Wind energy will expand to new regions worldwide, leveraging advancements in technology and accessibility to previously untapped wind resources.

Market Competitiveness: Cost reductions and technological advancements will make wind energy more competitive with conventional energy sources, driving further market penetration.

The future of wind turbine technology is poised for remarkable advancements, including larger and more efficient turbines, innovations in offshore wind, grid integration, and sustainability measures. Continued research and development will be instrumental in driving these advancements, ultimately contributing significantly to meeting renewable energy goals and transitioning towards a more sustainable and cleaner energy future.

Offshore Wind Farms: Expansion of offshore wind farms, utilizing higher capacity turbines, advanced installation techniques, and floating platforms, is becoming more prevalent due to the stronger and more consistent winds available at sea.

The expansion of offshore wind farms represents a significant trend in the renewable energy landscape, driven by advancements in technology and the favourable conditions available at sea. Here's a closer look at the key factors contributing to the growth of offshore wind farms:

Utilization of Higher Capacity Turbines:

Increased Energy Production: Deployment of larger and higher capacity turbines, with larger rotor diameters and greater power generation capabilities, leads to higher energy yields per turbine.

Enhanced Efficiency: Larger turbines can capture more wind energy at greater heights, taking advantage of stronger and more consistent winds available offshore.

Advanced Installation Techniques:

Specialized Vessels and Equipment: Dedicated vessels and specialized installation techniques cater to the unique challenges of installing offshore turbines in deeper waters and harsher marine environments.

Efficient Deployment: Innovations in installation processes and equipment aim to reduce construction times and costs, making offshore wind projects more commercially viable.

Floating Platforms:

Access to Deeper Waters: Floating offshore wind platforms enable access to deeper waters, where traditional fixed-bottom turbines may not be feasible, expanding potential installation sites.

Technological Advancements: Advancements in floating turbine technologies improve stability, making them suitable for various ocean conditions while capitalizing on stronger winds farther from shore.

Stronger and Consistent Winds at Sea:

Higher Wind Speeds: Offshore locations typically experience higher and more consistent wind speeds, leading to increased energy capture and more stable energy production.

Reduced Variability: Compared to onshore wind farms, offshore winds tend to be more consistent, contributing to a more reliable and steady power generation.

Environmental Impact Considerations:

Reduced Visual Impact: Offshore wind farms, when located far from shore, have less visual impact on coastal landscapes compared to onshore installations.

Minimized Noise Impact: Installation further offshore reduces the noise impact on nearby communities, contributing to improved acceptability.

Market Growth and Investment:

Growing Investment: Increasing investments and financial support from governments and private sectors are driving the growth of offshore wind projects worldwide.

Policy Support: Supportive regulatory frameworks and incentives encourage the development of offshore wind farms, fostering growth in the renewable energy sector.

Future Prospects:

Technology Refinement: Ongoing research and development aim to further improve offshore wind technology, making it more cost-effective and efficient.

Expansion into New Regions: Continued expansion of offshore wind farms into new geographic locations with untapped wind resources contributes to global renewable energy growth.

The expansion of offshore wind farms, facilitated by advancements in technology, larger turbines, advanced installation techniques, and floating platforms, offers substantial potential for harnessing clean and sustainable energy from wind resources available at sea. As these technologies continue to mature and costs decrease, offshore wind power is expected to play a significant role in meeting renewable energy targets and transitioning towards a more sustainable energy future.

Wind Energy Storage: Similar to solar, innovations in energy storage technologies support wind power by storing excess energy during peak generation for use when the wind is low.

energy storage technologies play a crucial role in supporting wind power by addressing intermittency issues and storing excess energy generated during periods of high wind for use when wind speeds are low or during peak demand. Here's a closer look at how energy storage complements wind energy:

Addressing Intermittency:

Storing Excess Energy: Energy storage systems, like batteries or pumped hydroelectric storage, capture surplus energy generated during high wind periods when demand is low.

Balancing Supply and Demand: Stored energy can be discharged during periods of low wind or high demand, ensuring a consistent and reliable supply of electricity.

Types of Energy Storage:

Battery Storage: Lithium-ion batteries and other advanced battery technologies store excess wind energy efficiently for later use.

Pumped Hydro Storage: This method uses surplus energy to pump water to an elevated reservoir and releases it through turbines to generate electricity when needed.

Flywheel Energy Storage: Rotating flywheels store kinetic energy and can release it rapidly when required, providing short-term energy support.

Grid Stability and Reliability:

Grid Support: Energy storage systems contribute to grid stability by providing backup power during fluctuations in wind energy output.

Frequency Regulation: Energy storage systems can respond rapidly to frequency changes, ensuring a stable grid when wind power output fluctuates.

Demand Management and Cost Savings:

Peak Shaving: Stored wind energy can be utilized during peak demand times, reducing reliance on expensive or polluting energy sources.

Optimized Use of Wind Energy: By storing excess energy, wind power's capacity to meet demand even during low wind periods or peak hours increases, maximizing its utilization.

Innovations in Storage Technologies:

Advancements in Battery Technology: Continued research aims to improve the efficiency, energy density, and longevity of batteries used for storing wind energy.

Grid-Scale Storage Solutions: Large-scale energy storage projects are being developed to support wind farms and increase grid reliability.

Future Prospects:

Cost Reduction: Ongoing advancements are expected to reduce the costs associated with energy storage technologies, making them more economically viable.

Integration with Renewables: Enhanced coordination between wind farms and energy storage systems will further optimize the use of renewable energy and grid stability.

Energy storage technologies play a vital role in complementing wind power by mitigating intermittency issues and ensuring a consistent and reliable energy supply. Continued innovation and integration of storage solutions with wind energy will further enhance grid stability, increase renewable energy penetration, and pave the way for a more sustainable and resilient energy future.

Hydro Energy

Pumped Hydro Storage: Advances in pumped hydro storage, a method of storing energy by pumping water uphill and releasing it downhill through turbines when electricity demand is high, continue to make it a viable solution for storing large amounts of energy.

pumped hydro storage is a mature and proven method for storing large amounts of energy by using water and gravitational potential energy. Advances in this technology continue to enhance its viability and effectiveness as a reliable energy storage solution. Here's a closer look at the advancements and benefits of pumped hydro storage:

Operating Principle:

Energy Storage Process: During periods of excess electricity generation or low demand, surplus energy is used to pump water from a lower reservoir to a higher elevation.

Electricity Generation: When demand increases or during peak hours, the stored water is released downhill, passing through turbines to generate electricity as it returns to the lower reservoir.

Advances and Benefits:

Efficiency and Reliability: Pumped hydro storage systems have high round-trip efficiency, typically over 70-80%, allowing for effective energy storage and retrieval.

Fast Response Time: These systems can respond quickly to shifts in demand, providing rapid injections of electricity to the grid when needed, enhancing grid stability.

Large-Scale Storage: Pumped hydro storage facilities can store massive amounts of energy, making them suitable for long-duration storage needs and supporting grid stability.

Long Lifecycle: These systems have a long lifespan, with minimal degradation over time, contributing to their reliability and cost-effectiveness.

Technological Advancements:

Advanced Turbines and Equipment: Upgrades in turbine technology and system components enhance efficiency and performance, contributing to higher energy yields.

Smart Management Systems: Integration of advanced control and monitoring systems optimizes the operation of pumped hydro facilities for better efficiency and response times.

Flexible Operation: Advancements allow for more flexible and dynamic operation, including rapid changes between pumping and generating modes based on grid demands.

Environmental Considerations:

Low Environmental Impact: Pumped hydro storage is considered environmentally friendly, utilizing water as the primary medium and having minimal emissions.

Location and Land Requirements: Suitable sites for pumped hydro storage require specific topographical features, but the land footprint can be minimal compared to other storage technologies.

Future Prospects:

Hybrid Solutions: Integration of pumped hydro storage with other energy storage technologies can enhance overall grid stability and flexibility.

Repurposing Abandoned Sites: Exploration of repurposing abandoned mines or quarries for pumped hydro storage facilities offers potential for sustainable energy storage expansion.

Economic Viability:

Cost Considerations: Continued advancements aim to reduce construction costs and improve the economic viability of pumped hydro storage projects.

Long-Term Investment: Despite higher upfront costs, the long lifecycle and reliability of pumped hydro storage make it a cost-effective option for large-scale energy storage.

Advancements in pumped hydro storage technology continue to solidify its position as a reliable and effective means of storing large amounts of energy. As research and development efforts persist, coupled with the exploration of innovative applications and repurposing opportunities, pumped hydro storage will play a significant role in supporting grid stability, enhancing renewable energy integration, and contributing to a more sustainable energy future.

Tidal and Wave Energy: Ongoing research explores harnessing energy from tides and waves. Various prototype technologies aim to efficiently capture energy from the ocean's movements.

The exploration of tidal and wave energy represents an ongoing field of research aimed at harnessing the kinetic and potential energy of ocean movements. Various prototype technologies are being developed to efficiently capture renewable energy from tidal currents and ocean waves. Here's an overview of these innovative approaches:

Tidal Energy:

Tidal Stream Generators: These devices resemble underwater wind turbines and are placed in tidal streams where water flow is strong. They capture kinetic energy from the ebb and flow of tides to generate electricity.

Barrage Systems: Tidal barrages involve constructing dams or barriers across estuaries or tidal rivers. As tides rise and fall, water passes through turbines, generating electricity.

Tidal Lagoons: Enclosed areas with tidal gates and turbines capture energy as water flows in and out of the lagoon during tidal changes, similar to a barrage system but in a confined area.

Wave Energy:

Point Absorber Devices: Floating buoys or devices on the ocean's surface capture energy from the vertical motion of waves, converting it into electricity.

Oscillating Water Columns (OWCs): These systems use the rise and fall of water levels within a chamber caused by wave motion to drive a column of air, which then spins a turbine to generate power.

Attenuators and Oscillating Wave Surge Converters (OWSCs): These long, floating structures or buoys capture energy from the movement of waves as they pass, converting it into electricity.

Research and Challenges:

Technological Innovation: Ongoing research focuses on improving the efficiency, reliability, and scalability of tidal and wave energy devices to make them commercially viable.

Environmental Impact: Designs aim to minimize environmental effects and disruption to marine ecosystems, ensuring sustainable deployment.

Durability and Maintenance: Challenges include designing devices that can withstand harsh marine conditions and reducing maintenance costs over time.

Advantages of Ocean Energy:

Renewable Nature: Tidal and wave energy sources are predictable, renewable, and less dependent on weather variations compared to some other renewable sources.

Low Carbon Footprint: Harnessing energy from tides and waves produces minimal greenhouse gas emissions, contributing to a cleaner energy mix.

Future Outlook:

Scaling Up Deployments: Continued advancements aim to scale up prototypes into larger arrays and commercial projects to increase energy production capacity.

Hybrid Energy Systems: Integration with other renewables or energy storage solutions to provide consistent and reliable electricity supply.

Research and development in tidal and wave energy technologies continue to progress, aiming to unlock the vast potential of the oceans as a renewable energy source. As innovations evolve and challenges are addressed, tidal and wave energy have the potential to play a significant role in diversifying the global renewable energy portfolio, contributing to sustainable and low-carbon energy generation.

Hydroelectric Efficiency: Improvements in the efficiency of traditional hydroelectric dams and turbines are being made to increase energy output and reduce environmental impacts.

Advancements in improving the efficiency of traditional hydroelectric dams and turbines are ongoing, aiming to enhance energy output while mitigating environmental impacts. Here are some key areas where improvements are being made:

Turbine Upgrades and Design Innovations:

Advanced Turbine Technology: Upgrading older turbines to more efficient designs, such as Kaplan, Francis, or Pelton turbines, improves energy conversion from flowing water to electricity.

Variable-Speed Turbines: Modernizing turbines with variable-speed technology allows for optimized performance under varying water flow conditions, increasing overall efficiency.

Fish-Friendly Turbines:

Fish Passage Designs: Innovative turbine designs include features that facilitate safe fish passage, reducing the impact on aquatic ecosystems and improving environmental sustainability.

Fish-Friendly Blade Shapes: Modifications to turbine blade shapes help prevent injury or mortality of fish passing through turbines, promoting aquatic biodiversity conservation.

Environmental Mitigation Measures:

Flow Management: Implementing controlled flow releases from dams based on environmental considerations helps maintain downstream ecosystems and habitats.

Sediment Management: Strategies to manage sediment buildup in reservoirs and downstream areas reduce ecological impacts on aquatic environments.

Efficiency Optimization and Retrofits:

Hydroelectric Plant Efficiency: Retrofitting existing hydroelectric facilities with modern equipment, controls, and automation systems improves overall plant efficiency and output.

Upgrading Infrastructure: Modernizing dam infrastructure, penstocks, and other components reduces energy losses and improves the performance of hydroelectric systems.

Computational Modelling and Simulation:

Hydrodynamic Studies: Advanced computational fluid dynamics (CFD) simulations aid in optimizing turbine designs and operational parameters for higher efficiency.

Virtual Testing: Simulation tools allow engineers to predict and refine the performance of upgraded turbines and dam configurations before physical implementation.

Benefits of Efficiency Improvements:

Increased Energy Output: Improved efficiency translates to higher electricity generation from existing hydroelectric facilities without significantly altering the environment.

Environmental Conservation: Reducing environmental impacts through fish-friendly designs and enhanced management practices supports aquatic ecosystems and biodiversity.

Future Prospects:

Hydropower Integration: Continued efforts to integrate advanced technologies and best practices will further optimize energy generation while minimizing environmental footprints.

Smart Grid Integration: Incorporating hydroelectric plants into smart grid systems enhances their flexibility and contributes to grid stability in the renewable energy mix.

Ongoing advancements in turbine technology, environmental considerations, and infrastructure improvements aim to enhance the efficiency of traditional hydroelectric dams. These efforts strive to increase energy output while minimizing environmental impacts, contributing to a more sustainable and efficient utilization of hydropower as a renewable energy source.

Other Renewables

Geothermal Energy: Developments in enhanced geothermal systems and new drilling technologies aim to make geothermal energy more accessible and cost-effective.

advancements in enhanced geothermal systems (EGS) and drilling technologies are revolutionizing the accessibility and cost-effectiveness of geothermal energy, tapping into the Earth's heat to produce renewable and sustainable electricity. Here are key developments in this field:

Enhanced Geothermal Systems (EGS):

Engineering Reservoirs: EGS involves creating or stimulating geothermal reservoirs by injecting water into hot rocks, creating fractures that enhance heat transfer for electricity generation.

Improved Heat Extraction: Enhanced circulation techniques, such as hydraulic fracturing (or "fracking") and thermal stimulation, aim to increase permeability and heat extraction rates.

Advanced Drilling Technologies:

Directional Drilling: Utilizing advanced drilling techniques allows for horizontal and directional drilling, increasing access to deeper and hotter geothermal resources.

Slim Hole Drilling: Smaller, more efficient drilling techniques reduce costs and environmental impact while reaching suitable geothermal reservoirs.

Innovative Heat Extraction Methods:

Binary Cycle Technology: Advanced binary cycle systems use lower-temperature geothermal resources by vaporizing a fluid with a lower boiling point than water to generate electricity.

Closed-Loop Systems: Systems that circulate a working fluid (like a refrigerant) in a closed-loop to capture geothermal heat and produce electricity more efficiently.

Reservoir Engineering and Modelling:

Geophysical Surveys: Improved imaging and modelling techniques help identify and characterize geothermal resources more accurately, aiding in optimal reservoir placement.

Predictive Models: Advanced modelling and simulation tools predict the behaviour of geothermal reservoirs, aiding in optimal resource management and energy extraction.

Benefits and Advantages:

Reliable Baseload Power: Geothermal energy provides continuous, baseload power, offering a stable and predictable renewable energy source.

Low Carbon Footprint: Geothermal energy production emits minimal greenhouse gases, contributing to cleaner energy generation and reduced environmental impact.

Future Outlook:

Cost Reduction: Continued research and technology development aim to reduce upfront costs associated with exploration, drilling, and reservoir development.

Market Expansion: Advancements will likely drive increased adoption of geothermal energy in regions previously considered unsuitable or less explored.

Environmental Considerations:

Minimal Land Use Impact: Geothermal plants have relatively small land footprints, reducing the impact on surrounding ecosystems compared to some other energy sources.

Mitigation of Emissions: Continued advancements aim to reduce or eliminate emissions associated with drilling and energy extraction processes.

Developments in enhanced geothermal systems and drilling technologies are pivotal in expanding the accessibility and cost-effectiveness of geothermal energy. These innovations offer a promising path toward unlocking more of the Earth's renewable heat resources, contributing to a sustainable and clean energy future.

Bioenergy Innovations: Advancements in bioenergy focus on optimizing the conversion of organic materials, such as agricultural waste or algae, into usable energy sources like biofuels and biogas.

Advancements in bioenergy technologies are revolutionizing the conversion of organic materials into usable energy sources like biofuels and biogas. Here are some key innovations and developments in this field:

Biofuels from Agricultural Waste:

Advanced Feedstock Processing: Improved methods for breaking down lignocellulosic materials (such as agricultural residues) into sugars for biofuel production.

Biochemical Conversion: Utilizing enzymes and microorganisms to ferment sugars from cellulose and hemicellulose into biofuels like ethanol and biodiesel.

Algae Biofuels:

Algae Cultivation Technology: Innovations in algae growth systems, such as photobioreactors and open pond systems, to enhance productivity and lipid content for biofuel production.

Genetic Engineering: Research into genetic modification of algae strains to improve oil content, growth rates, and resistance to environmental stressors.

Research in genetic engineering of algae strains aims to enhance their traits, such as oil content, growth rates, and resilience to environmental stressors, making them more efficient for biofuel production. Here are key aspects of this research:

Improving Oil Content:

Manipulating Lipid Biosynthesis: Genetic modifications target pathways involved in lipid production within algae cells, increasing the accumulation of oils suitable for biofuel extraction.

Enhancing Oil Quality: Altering specific genetic factors can lead to the production of oils with better fuel properties, such as higher energy density or improved stability.

Increasing Growth Rates:

Enhanced Photosynthesis: Genetic engineering aims to optimize photosynthetic pathways, allowing algae to capture and convert sunlight more efficiently, boosting growth rates.

Nutrient Uptake Enhancement: Genetic modifications can improve the uptake and utilization of nutrients, promoting faster growth and higher biomass yields.

Environmental Stress Resistance:

Temperature Tolerance: Algae strains are modified to withstand a broader range of temperatures, enabling cultivation in diverse environments without compromising growth rates.

Salinity and pH Tolerance: Genetic modifications enhance algae resilience to varying salinity levels and pH fluctuations in water, increasing their adaptability to different conditions.

Precision Gene Editing Techniques:

CRISPR/Cas9 Technology: Advanced gene editing tools like CRISPR/Cas9 enable precise modifications in the algae genome, facilitating targeted alterations for desired traits.

Transgenic Approaches: Introducing genes from other organisms into algae genomes can confer specific advantageous characteristics, such as stress tolerance or enhanced productivity.

Regulation and Ethics:

Environmental Safety Considerations: Research takes into account potential ecological impacts and implements measures to prevent unintended consequences of modified algae in natural ecosystems.

Ethical Guidelines: Adherence to ethical guidelines and regulations governing genetic engineering ensures responsible and safe research practices.

Future Prospects:

Commercial Viability: Success in enhancing traits through genetic engineering could lead to the development of more commercially viable algae strains for biofuel production.

Sustainable Biofuel Production: Genetically optimized algae strains have the potential to contribute significantly to sustainable and renewable biofuel production.

Research in genetic engineering of algae strains focuses on enhancing their oil content, growth rates, and resilience to environmental stressors. These advancements aim to create more efficient and adaptable algae species suitable for large-scale biofuel production, contributing to the shift towards cleaner and renewable energy sources. Ethical considerations and regulatory oversight remain crucial to ensure the safe and responsible application of genetic engineering in algae research for bioenergy.

Biogas Production from Organic Waste:

Anaerobic Digestion: Improved anaerobic digestion processes to efficiently break down organic matter like food waste, agricultural residues, and sewage sludge to produce biogas.

anaerobic digestion is a biological process that efficiently breaks down organic matter, including food waste, agricultural residues, sewage sludge, and other organic materials, to produce biogas. Advancements in anaerobic digestion processes have led to more efficient and sustainable biogas production. Here are key improvements:

Process Optimization:

Optimal Temperature and pH Control: Improved anaerobic digestion systems maintain specific temperature and pH conditions to maximize microbial activity and biogas yield.

Mixing and Homogenization: Enhanced mixing and agitation mechanisms ensure efficient digestion by evenly distributing organic materials and microorganisms throughout the digester.

Advanced Digestion Technologies:

High-Solids Digestion: Innovations allow anaerobic digestion of higher solids content, enabling the processing of a wider range of organic wastes and increasing overall efficiency.

Continuous Flow Systems: Continuous systems enhance biogas production by allowing a continuous feed of organic materials into the digester, maintaining a stable microbial environment.

Co-digestion and Pre-treatment:

Co-digestion of Multiple Feedstocks: Combining various organic wastes, such as food waste, crop residues, and animal manure, optimizes digestion and enhances biogas production.

Pre-treatment Methods: Pre-treatment technologies like shredding, grinding, or enzymatic treatment break down complex materials, improving digestion efficiency.

Biogas Quality Enhancement:

Biogas Upgrading: Advanced purification technologies remove impurities like CO_2, moisture, and hydrogen sulfide from biogas, increasing its energy content and quality.

Digestate Management: Improved separation and treatment of digestate (the residue after digestion) reduce environmental impact and enhance its value as a nutrient-rich fertilizer.

Automation and Control Systems:

Monitoring and Control: Advanced sensors and control systems regulate key parameters like temperature, pH, and gas production, optimizing the digestion process.

Remote Monitoring: Remote monitoring and automation technologies allow operators to oversee and manage digesters from a distance, improving operational efficiency.

Environmental and Economic Benefits:

Renewable Energy Generation: Biogas produced from anaerobic digestion is a renewable energy source that can be used for heat, electricity, or vehicle fuel.

Waste Diversion and Nutrient Recovery: Anaerobic digestion diverts organic waste from landfills, reducing methane emissions, and produces digestate as a valuable fertilizer.

Future Prospects:

Scaling Up and Integration: Larger-scale digestion facilities and integration with other renewable energy systems enhance biogas production and contribute to a more circular economy.

Technological Innovation: Continued research focuses on developing more efficient digestion processes and exploring novel feedstocks to further optimize biogas production.

Improved anaerobic digestion processes efficiently break down organic matter to produce biogas, offering renewable energy generation and waste diversion benefits. Advancements in technology and process optimization continue to enhance biogas production from various organic sources, contributing to sustainable waste management and renewable energy goals.

Co-digestion and Pre-treatment: Mixing various organic materials for co-digestion and employing pre-treatment methods to enhance biogas yield and quality.

Co-digestion and pre-treatment are essential processes in anaerobic digestion, optimizing the breakdown of organic materials to enhance biogas yield and quality. Here's a detailed look at these methods:

Co-digestion:

Diverse Feedstock Utilization: Co-digestion involves combining different organic materials, such as food waste, crop residues, manure, and wastewater sludge, in the anaerobic digester.

Synergistic Effects: Mixing various feedstocks can create a more balanced nutrient composition, providing a more favourable environment for microorganisms, thus enhancing biogas production.

Balancing Carbon-to-Nitrogen Ratio: Combining materials with varied C/N ratios helps achieve an optimal balance for efficient microbial activity and gas production.

Pre-treatment:

Mechanical Processing: Shredding, grinding, or chopping organic materials increases surface area, facilitating microbial access and accelerating digestion.

Chemical and Enzymatic Treatment: Chemical additives or enzyme applications break down complex compounds, making them more accessible to microbial degradation.

Thermal Treatment: Heat treatment or pasteurization kills pathogens, improves digestibility, and enhances the breakdown of organic matter.

Benefits of Co-digestion and Pre-treatment:

Enhanced Biogas Yield: Co-digestion broadens the feedstock range, providing a wider spectrum of nutrients for microbes, resulting in increased biogas production.

Improved Digestion Efficiency: Pre-treatment methods break down complex compounds, making organic materials more digestible, leading to faster and more complete digestion.

Process Stability: Co-digestion balances nutrient ratios, reducing process instability and improving the robustness of the anaerobic digestion system.

Optimization Strategies:

Feedstock Selection: Identifying complementary feedstocks that optimize nutrient balance and improve digestion efficiency is crucial for successful co-digestion.

Pre-treatment Specificity: Selecting appropriate pre-treatment methods tailored to the feedstock characteristics optimizes efficiency and enhances biogas yield.

Monitoring and Control: Continuous monitoring of digestion parameters helps adjust the process, ensuring optimal conditions for microbial activity.

Environmental and Economic Impacts:

Waste Diversion: Co-digestion diverts various organic wastes from landfills, reducing methane emissions and promoting sustainable waste management.

Resource Recovery: Biogas production from co-digestion generates renewable energy, while the resulting digestate can be used as a nutrient-rich fertilizer.

Future Developments:

Integration with Circular Economy: Co-digestion aligns with circular economy principles by valorising diverse organic wastes into renewable energy and valuable by-products.

Technological Innovations: Continued research focuses on innovative pre-treatment methods and feedstock combinations to further optimize biogas production.

Co-digestion of diverse organic materials and pre-treatment methods are pivotal strategies in anaerobic digestion, optimizing biogas production, improving process stability, and contributing to sustainable waste management while generating renewable energy resources. Continued advancements and optimized approaches in these methods will further drive efficiency and utilization in the bioenergy sector.

Bioenergy Integration and Hybrid Systems:

Combined Heat and Power (CHP): Integration of bioenergy systems with heat and power generation to maximize energy utilization and overall efficiency.

Combined Heat and Power (CHP), also known as cogeneration, is a highly efficient energy generation process that integrates bioenergy systems, simultaneously producing usable heat and electricity. Here's an overview of CHP and its integration with bioenergy systems:

CHP Process:

Simultaneous Energy Generation: CHP systems generate electricity while also capturing and utilizing the excess heat produced during electricity generation.

Utilization of Waste Heat: Heat that would otherwise be wasted in conventional power generation processes is harnessed for various purposes, such as heating buildings or industrial processes.

Integration with Bioenergy Systems:

Biomass CHP: Biomass-based systems, like biomass boilers or biogas plants, can be integrated into CHP systems to produce both electricity and heat.

Synergy with Anaerobic Digestion: Biogas produced from anaerobic digestion of organic waste can fuel CHP units, generating electricity and using the heat for heating purposes.

Benefits of CHP Integration with Bioenergy:

Energy Efficiency: CHP systems typically reach higher overall efficiencies compared to separate electricity and heat generation, utilizing a higher percentage of the fuel's energy content.

Cost Savings: By capturing and utilizing waste heat, CHP systems reduce the need for separate heating sources, leading to cost savings in energy bills.

Reduced Emissions: Increased efficiency in energy conversion leads to reduced greenhouse gas emissions compared to conventional power generation methods.

Applications and Sectors:

Industrial Settings: CHP is commonly utilized in industries with high heat and electricity demand, such as manufacturing, refining, and food processing.

Commercial Buildings: Hospitals, universities, hotels, and residential complexes can benefit from CHP systems for heating, cooling, and electricity needs.

Challenges and Considerations:

System Sizing: Properly sizing CHP units to match both electrical and thermal demands is crucial for maximizing efficiency.

Regulatory and Financial Hurdles: Regulatory barriers and financial incentives can impact the adoption and implementation of CHP systems.

Future Trends:

Smart Grid Integration: Integration with smart grid technologies allows CHP systems to operate flexibly and efficiently, adapting to variable energy demands.

Renewable Fuel Integration: Advancements in using renewable biofuels in CHP systems contribute to a more sustainable and greener energy mix.

CHP systems integrated with bioenergy offer a highly efficient and sustainable approach to energy generation. By harnessing both electricity and heat from bioenergy sources like biomass or biogas, CHP maximizes energy utilization, reduces emissions, and provides cost-effective solutions for diverse sectors, contributing to a more sustainable and resilient energy infrastructure.

Synergies with Other Renewables: Exploring hybrid systems that combine bioenergy with solar, wind, or geothermal sources for a more balanced and reliable energy supply.

exploring hybrid systems that combine bioenergy with other renewable sources like solar, wind, or geothermal energy offers promising avenues for a more balanced, reliable, and sustainable energy supply. Here's an overview of these synergies:

Bioenergy-Solar Hybrid Systems:

Complementary Nature: Solar energy production is often highest during daylight hours, while bioenergy can provide a continuous baseload supply. Combining these sources creates a more stable energy output.

Shared Infrastructure: Integrated systems can share grid connections and some infrastructure components, optimizing costs and land use.

Bioenergy-Wind Hybrid Systems:

Balanced Energy Generation: Wind power tends to be more variable, while bioenergy offers stability. Combining these sources provides a more consistent energy supply.

Land Use Optimization: Wind turbines and bioenergy facilities can share land, making efficient use of available space for energy production.

Bioenergy-Geothermal Hybrid Systems:

Stability and Reliability: Geothermal energy provides a constant baseload, while bioenergy can offer flexibility. Combined systems ensure a steady and reliable energy supply.

Operational Synergy: Integrated operations and management between bioenergy and geothermal facilities can optimize energy output and infrastructure utilization.

Benefits of Hybrid Systems:

Increased Reliability: Combining different renewable sources smoothens fluctuations in energy output, offering a more stable supply to the grid.

Enhanced Grid Integration: Hybrid systems can provide more consistent energy, facilitating better integration into existing power grids.

Technological Challenges:

System Integration: Designing and integrating different technologies and managing their interactions within hybrid systems require advanced engineering and control systems.

Resource Matching: Aligning the varying availability of different renewable sources requires careful planning and management.

Future Outlook:

Technological Advancements: Continued research and development aim to improve hybrid system designs and optimize their operation for increased efficiency.

Market Adoption: Increasing interest in diverse and integrated energy solutions is expected to drive the adoption of bioenergy hybrid systems.

Environmental Impact:

Reduced Emissions: Hybrid systems contribute to lower greenhouse gas emissions by combining renewable sources and reducing reliance on fossil fuels.

Sustainable Energy Mix: A diversified energy mix derived from multiple renewables supports a sustainable and resilient energy future.

Exploring hybrid systems that combine bioenergy with other renewable sources offers a pathway to a more balanced, reliable, and sustainable energy supply. These integrated systems, combining the strengths of different renewables, play a crucial role in diversifying the energy mix and fostering a cleaner and more resilient energy infrastructure. Continued innovation and integration will further enhance the viability and efficiency of such hybrid systems in the future.

Waste-to-Energy Innovations:

Bioenergy from Waste Streams: Conversion of various organic waste streams, including agricultural residues, forestry waste, and municipal solid waste, into energy.

Bioenergy derived from waste streams involves converting various organic materials, such as agricultural residues, forestry waste, and municipal solid waste (MSW), into energy sources like biogas, biofuels, or heat and electricity. Here's an overview of bioenergy production from these waste streams:

Biomass-to-Energy Conversion:

Anaerobic Digestion: Organic waste undergoes anaerobic digestion, breaking down biodegradable materials in the absence of oxygen to produce biogas, primarily methane and carbon dioxide.

Biogas Utilization: Biogas produced from organic waste can be used directly for heating, electricity generation, or upgraded to biomethane for vehicle fuel or injection into the natural gas grid.

Agricultural Residues:

Crop Residues: Byproducts from agricultural activities, such as corn stover, rice husks, and wheat straw, can be processed into biofuels like ethanol or utilized in biogas production.

Animal Manure: Livestock waste, like cow or pig manure, can undergo anaerobic digestion to produce biogas and nutrient-rich digestate for use as fertilizer.

Forestry Waste:

Wood Residues: Waste wood, forest residues, or sawdust can be converted into bioenergy through processes like combustion, gasification, or pyrolysis.

Black Liquor: Waste product from the pulp and paper industry can be used in combined heat and power (CHP) systems to generate both heat and electricity.

Municipal Solid Waste (MSW):

Waste-to-Energy (WtE) Plants: MSW can be incinerated in WtE plants to generate heat or electricity, reducing the volume of waste while producing energy.

Landfill Gas Recovery: Landfills capture methane produced by decomposing organic waste, utilizing it for electricity generation or as a fuel source.

Benefits of Bioenergy from Waste Streams:

Waste Reduction: Conversion of organic waste into energy reduces landfill volumes and greenhouse gas emissions from decomposing waste.

Renewable Energy Generation: Bioenergy derived from waste sources contributes to renewable energy portfolios, decreasing reliance on fossil fuels.

Challenges and Considerations:

Feedstock Availability: Ensuring a consistent and reliable supply of waste feedstock for bioenergy production can be a logistical challenge.

Technology Optimization: Improving conversion technologies and optimizing processes for different waste streams to increase energy output and efficiency.

Future Developments:

Technological Innovations: Research focuses on enhancing waste-to-energy conversion technologies and developing new methods for more efficient resource utilization.

Circular Economy Integration: Continued efforts aim to align bioenergy production from waste streams with circular economy principles, promoting resource recovery and reuse.

Bioenergy derived from organic waste streams presents a valuable opportunity to convert waste materials into useful energy sources while contributing to waste reduction and renewable energy production. Continued advancements in technology and sustainable waste management practices will further enhance the potential and efficiency of bioenergy generation from various waste sources, supporting a more sustainable and resource-efficient future.

Efficiency Improvements: Advancements in gasification, pyrolysis, and combustion technologies to extract energy more efficiently from diverse waste materials.

Advancements in gasification, pyrolysis, and combustion technologies play a crucial role in improving the efficiency of extracting energy from diverse waste materials. Here's an overview of these technologies and their efficiency improvements:

Gasification:

Process Enhancement: Gasification converts organic materials into synthesis gas (syngas) using controlled amounts of oxygen or steam in a high-temperature environment, allowing for cleaner and more efficient energy extraction.

Syngas Utilization: Syngas, composed of carbon monoxide, hydrogen, and other gases, can be used in combustion engines, turbines, or fuel cells for electricity or heat generation.

Efficiency Enhancement: Advancements aim to optimize gasifier designs, enhance gas cleaning methods, and improve gas-to-energy conversion technologies for higher overall system efficiency.

Pyrolysis:

Thermal Decomposition: Pyrolysis involves heating organic material in the absence of oxygen to produce bio-oil, biochar, and syngas, offering a method to convert various waste materials into valuable products.

Bio-Oil Utilization: Bio-oil obtained from pyrolysis can be used as a renewable fuel or feedstock for producing chemicals or as a substitute for fossil fuels.

Technological Refinement: Research focuses on enhancing pyrolysis reactor designs, optimizing heating rates, and improving feedstock pre-processing for increased energy recovery and process efficiency.

Combustion:

Heat Generation: Combustion involves burning waste materials to produce heat, which can be used directly in industrial processes or to generate electricity in boilers or steam turbines.

Efficiency Enhancements: Advanced combustion technologies, including fluidized bed combustion or co-firing in power plants, aim to increase efficiency and reduce emissions from waste combustion.

Emission Control: Integration of emission control technologies, such as scrubbers or filters, helps minimize pollutants released during combustion, ensuring environmental compliance.

Benefits of Efficiency Improvements:

Higher Energy Recovery: Enhanced technologies allow for better extraction of energy from waste materials, increasing overall energy output and resource utilization.

Reduced Environmental Impact: Improved efficiency in waste-to-energy processes leads to reduced emissions, helping mitigate environmental pollution and climate impacts.

Challenges and Considerations:

Feedstock Variability: Different waste materials pose challenges in terms of consistency and composition, requiring adaptable technologies for efficient conversion.

Economic Viability: Balancing technological advancements with cost-effectiveness remains crucial for the widespread adoption of efficient waste-to-energy systems.

Future Outlook:

Integration of Technologies: Hybrid approaches combining gasification, pyrolysis, and combustion technologies could offer more efficient and versatile waste-to-energy solutions.

Continued R&D: Ongoing research focuses on scaling up and refining these technologies to improve energy recovery rates, reduce costs, and enhance environmental sustainability.

Advancements in gasification, pyrolysis, and combustion technologies are vital for improving the efficiency of extracting energy from diverse waste materials. These innovations contribute to increasing energy recovery rates, reducing environmental impact, and fostering the development of more sustainable and efficient waste-to-energy solutions for a cleaner and greener future.

Sustainability and Environmental Considerations:

Carbon Neutrality: Bioenergy from organic sources is considered carbon-neutral as it utilizes CO_2 already present in the carbon cycle, reducing net greenhouse gas emissions.

Bioenergy derived from organic sources, such as biomass, agricultural residues, forestry waste, or biogas from organic waste, is often regarded as carbon-neutral because it primarily recycles carbon that is already part of the natural carbon cycle. Here's an explanation of why bioenergy is considered carbon-neutral:

Carbon Neutrality in Bioenergy:

Closed Carbon Cycle: Biomass used for bioenergy purposes absorbs carbon dioxide (CO_2) from the atmosphere during its growth through photosynthesis. When this biomass is converted into energy, it releases CO_2 back into the atmosphere.

Neutral Carbon Impact: The CO_2 released during combustion or degradation of organic materials for bioenergy is considered part of the natural carbon cycle. It doesn't introduce new carbon into the atmosphere but recycles carbon that was recently captured from the atmosphere.

Benefits and Considerations:

Emission Offset: Bioenergy contributes to offsetting greenhouse gas emissions by recycling CO_2 that would be released naturally as the organic matter decomposes or burns.

Reduction in Fossil Fuel Use: Using bioenergy as a renewable energy source displaces fossil fuel consumption, reducing overall CO_2 emissions and dependency on non-renewable resources.

Lifecycle Analysis and Impacts:

Lifecycle Assessments: Evaluating the overall environmental impact of bioenergy includes considering emissions from cultivation, harvesting, processing, and transportation of biomass.

Land Use and Sustainability: Ensuring sustainable practices in biomass production is crucial to maintaining its carbon-neutral status, avoiding deforestation or land-use changes that could release additional CO_2.

Challenges and Criticisms:

Time Lag: The carbon neutrality of bioenergy assumes immediate re-absorption of CO_2 by new biomass growth. However, this may take time, and the timing of emissions versus reabsorption needs consideration.

Biomass Source and Efficiency: The carbon neutrality of bioenergy heavily depends on the type of biomass used, conversion technology efficiency, and sustainable management practices.

Future Prospects:

Technological Advancements: Improving conversion efficiencies and developing sustainable biomass sourcing practices can enhance the carbon neutrality of bioenergy.

Policy and Certification: Regulatory frameworks and certifications can ensure sustainable bioenergy production, addressing concerns and maintaining carbon-neutral status.

Bioenergy derived from organic sources is often considered carbon-neutral due to its reliance on the carbon cycle. However, ensuring true carbon neutrality requires sustainable sourcing, efficient conversion processes, and accurate lifecycle assessments. Bioenergy plays a role in reducing net greenhouse gas emissions and transitioning towards a more sustainable and low-carbon energy future when managed responsibly and sustainably.

Waste Reduction and Circular Economy: Bioenergy systems contribute to waste reduction by repurposing organic materials into valuable energy sources, aligning with circular economy principles.

bioenergy systems play a significant role in waste reduction and align with the principles of a circular economy by repurposing organic materials into valuable energy sources. Here's how bioenergy contributes to waste reduction within the framework of a circular economy:

Organic Waste Valorization:

Utilization of Biomass: Bioenergy systems utilize various forms of organic waste, including agricultural residues, forestry waste, food scraps, and animal manure, which would otherwise be disposed of in landfills or left to decompose.

Conversion to Energy: These organic materials are processed in bioenergy systems (such as anaerobic digestion, gasification, or combustion) to produce biogas, biofuels, heat, or electricity, thereby reducing waste volumes and converting them into valuable energy.

CIRCULAR ECONOMY PRINCIPLES:

Resource Regeneration: Bioenergy systems contribute to regenerating resources by turning organic waste into usable energy, essentially closing the loop on waste materials and reusing them as a resource.

Waste Minimization: Instead of landfilling or incinerating organic waste, bioenergy processes minimize waste by extracting energy value, reducing the environmental impact of waste disposal.

Benefits of Waste Valorization through Bioenergy:

Reduced Landfill Pressure: Bioenergy systems divert organic waste from landfills, decreasing methane emissions from decomposing waste and reducing the need for additional landfill space.

Green Energy Generation: By converting organic waste into renewable energy sources, bioenergy systems contribute to cleaner energy production, reducing reliance on non-renewable resources.

Circular Economy Integration:

Resource Efficiency: Bioenergy aligns with the principles of resource efficiency by extracting energy from organic waste, ensuring a more efficient use of resources within the system.

Sustainable Waste Management: Integrating bioenergy into waste management strategies fosters a more sustainable approach by valorizing waste and reducing its environmental impact.

Challenges and Considerations:

Feedstock Availability: Ensuring a consistent supply of diverse organic waste streams for bioenergy production can be a logistical challenge.

Technology Optimization: Enhancing bioenergy technologies and processes to efficiently handle various waste types and improve overall energy recovery is crucial.

Future Prospects:

Policy Support: Policies promoting waste valorization and renewable energy can further incentivize the integration of bioenergy into circular economy strategies.

Technological Innovation: Continued advancements in bioenergy technologies and sustainable waste management practices will optimize waste valorization and energy production.

Bioenergy systems contribute significantly to waste reduction by repurposing organic materials into valuable energy sources, aligning with circular economy principles of resource regeneration and waste minimization. By utilizing organic waste as a resource for energy production, bioenergy plays a vital role in sustainable waste management and renewable energy generation, fostering a more circular and resource-efficient economy.

Future Prospects:

Technological Refinement: Continued research aims to improve conversion efficiencies, reduce costs, and develop more sustainable and scalable bioenergy production systems.

continued research and development efforts are pivotal for enhancing bioenergy production systems to improve efficiency, reduce costs, and ensure sustainability. Here's an overview of ongoing technological refinement in bioenergy:

Improved Conversion Efficiencies:

Enhanced Biomass-to-Energy Processes: Research focuses on optimizing conversion technologies such as gasification, pyrolysis, anaerobic digestion, and combustion to increase energy output and overall system efficiency.

Innovative Reactor Designs: Development of advanced reactor designs and process optimization techniques to maximize energy recovery from diverse biomass feedstocks.

Cost Reduction:

Economies of Scale: Scaling up bioenergy production systems to larger capacities helps reduce production costs per unit of energy by benefiting from economies of scale.

Efficiency Upgrades: Innovations in equipment, automation, and process improvements aim to reduce operational costs and increase overall energy output.

Sustainable Bioenergy Systems:

Feedstock Diversification: Research explores utilizing a wider range of feedstocks, including agricultural residues, algae, dedicated energy crops, and organic waste, ensuring a diverse and sustainable resource base.

Waste Valorization: Technologies are being refined to extract energy efficiently from various waste streams, contributing to waste reduction and renewable energy production.

Advanced Technologies:

Biorefineries: Integration of various bioenergy processes within biorefineries to produce a range of bio-based products, including fuels, chemicals, and materials, maximizing resource utilization.

Next-Generation Feedstock Pretreatment: Innovative pre-treatment methods improve the breakdown of complex biomass materials, enhancing bioenergy yields and process efficiencies.

Sustainability Assessment:

Lifecycle Analysis: Continued research evaluates the environmental and social impacts of bioenergy systems throughout their lifecycle, ensuring sustainability and mitigating potential drawbacks.

Carbon Sequestration: Exploring methods to capture and sequester carbon from bioenergy processes, maintaining a carbon-neutral or carbon-negative balance.

Integration and Interdisciplinary Approaches:

Cross-Sector Collaboration: Collaborative efforts among researchers, industry, and policymakers facilitate knowledge sharing and accelerate technological advancements.

Interdisciplinary Research: Integration of engineering, biology, environmental sciences, and economics fosters holistic approaches to enhance bioenergy systems.

Future Prospects:

Techno-Economic Analysis: Continued techno-economic assessments to identify the most cost-effective and sustainable bioenergy pathways for deployment.

Policy Support and Innovation: Supportive policies and funding initiatives drive innovation and incentivize the adoption of sustainable bioenergy technologies.

Continued research and development in bioenergy technologies are crucial for enhancing conversion efficiencies, reducing costs, and ensuring the sustainability and scalability of bioenergy production. These advancements pave the way for a more efficient, cost-effective, and environmentally friendly bioenergy sector, contributing significantly to renewable energy portfolios and sustainable resource management.

Market Expansion: Bioenergy is expected to play a significant role in meeting renewable energy targets and fostering a more diversified and sustainable energy portfolio globally.

Bioenergy is poised to play a substantial role in meeting renewable energy targets and contributing to a more diversified and sustainable global energy portfolio. Here's why:

Renewable Energy Contribution:

Diversification of Energy Sources: Bioenergy provides a versatile renewable energy source that complements other intermittent renewables like wind and solar, contributing to a more balanced energy mix.

Baseload and Dispatchable Energy: Certain bioenergy forms, such as biogas or biomass power, offer baseload or dispatchable power, enhancing grid stability and reliability.

Global Energy Demand:

Meeting Energy Demand: As global energy demands continue to rise, bioenergy's potential to supply sustainable energy from diverse feedstocks becomes increasingly crucial.

Energy Security: Bioenergy reduces dependence on finite fossil fuel reserves, enhancing energy security by utilizing locally available biomass resources.

Environmental and Climate Benefits:

Greenhouse Gas Mitigation: Bioenergy contributes to mitigating greenhouse gas emissions, particularly when derived from sustainably managed biomass, offering a carbon-neutral or even carbon-negative energy option.

Waste Reduction: Bioenergy systems help manage organic waste, minimizing environmental impacts associated with landfilling and methane emissions.

Rural Development and Employment:

Rural Economies: Biomass production for bioenergy can stimulate rural economies by creating jobs in agriculture, forestry, and the bioenergy sector.

Local Resource Utilization: Utilizing locally available biomass resources for energy production can enhance local economic development and resource utilization.

Market Expansion Initiatives:

Technology Standardization: Efforts to standardize and improve bioenergy technologies ensure reliability, scalability, and broader market acceptance.

Policy Support: Supportive policies, incentives, and regulations at national and international levels encourage investment in bioenergy projects and market expansion.

Innovation and Investment:

Research and Development: Ongoing innovation drives efficiency improvements, cost reductions, and the development of advanced bioenergy technologies.

Private and Public Investment: Increased investment in bioenergy projects, both from public and private sectors, fuels technological advancements and market expansion.

Global Energy Transition:

Renewable Energy Targets: Bioenergy contributes to achieving national and international renewable energy goals outlined in climate agreements, such as the Paris Agreement.

Transition to Sustainable Energy: Bioenergy facilitates the transition to a more sustainable energy landscape by providing a reliable and renewable energy source.

Bioenergy's versatility, renewable nature, and potential to mitigate climate change make it a pivotal component in meeting renewable energy targets globally. As technological advancements, supportive policies, and increased investment continue to drive market expansion, bioenergy will play an integral role in fostering a more diversified, sustainable, and resilient global energy portfolio.

Advancements in bioenergy technologies are driving the efficient conversion of organic materials into valuable energy sources like biofuels and biogas. These innovations contribute to waste reduction, promote sustainability, and offer promising solutions in the transition towards cleaner and more sustainable energy production.

Hybrid Systems: Integration of multiple renewable energy sources into hybrid systems allows for more reliable and continuous power generation. Combinations like solar-wind, wind-hydro, etc., aim to complement each other's fluctuations and enhance overall efficiency.

The integration of multiple renewable energy sources into hybrid systems offers several advantages by combining the strengths of each source to ensure more reliable and continuous power generation. Here's an overview of the benefits and examples of hybrid renewable energy systems:

Advantages of Hybrid Systems:

Enhanced Reliability: Combining different renewables mitigates the intermittency issue of individual sources, ensuring a more consistent and reliable energy supply.

Optimized Energy Output: Hybrid systems leverage complementary characteristics of various renewables, maximizing overall energy generation potential.

Examples of Hybrid Renewable Energy Systems:

Solar-Wind Hybrid Systems: Integrating solar photovoltaic (PV) panels with wind turbines enables energy generation during different times of the day and seasons, compensating for variations in weather conditions.

Wind-Hydro Hybrid Systems: Combining wind power with hydroelectricity offers stability as wind generation can be unpredictable, while hydropower provides a steady baseload or peak power.

Solar-Hydro Hybrid Systems: Pairing solar power with hydroelectricity allows for continuous power generation as solar energy peaks during the day while hydropower offers consistent energy.

Wind-Solar-Battery Hybrid Systems: Adding battery storage to wind-solar hybrids enables the storage of excess energy during peak production for use during low generation periods, ensuring a more constant power supply.

Operational Synergies:

Complementary Nature: Different renewable sources have varying generation patterns, so combining them can balance energy generation and utilization.

Resource Utilization: Leveraging diverse renewables optimizes resource utilization, tapping into multiple energy sources available at different times.

Challenges and Considerations:

Integration Complexity: Designing hybrid systems requires sophisticated planning to synchronize multiple energy sources efficiently.

Optimal Sizing: Properly sizing each component (solar panels, wind turbines, storage systems) to match energy demand is crucial for maximizing system efficiency.

Future Prospects:

Technological Advances: Continued advancements in energy storage, smart grid technologies, and control systems will further optimize hybrid systems.

Market Adoption: Increasing interest in hybrid systems due to their reliability and efficiency is expected to drive wider adoption and deployment.

Hybrid renewable energy systems, integrating multiple sources such as solar, wind, hydro, or storage technologies, offer a promising solution to ensure a reliable and continuous energy supply. By combining the strengths of different renewables, these systems contribute to increased energy reliability, reduced intermittency, and enhanced overall efficiency, playing a crucial role in a more sustainable and resilient energy future.

Energy Storage Solutions

Solid-State Batteries: This technology aims to replace traditional lithium-ion batteries with more stable and higher-energy-density alternatives. Solid-state batteries have the potential to increase energy storage capacity, improve safety, and reduce charging times.

Solid-state batteries represent a promising advancement in energy storage technology with the potential to surpass traditional lithium-ion batteries in various aspects. Here's an overview of solid-state batteries and their potential advantages:

Characteristics of Solid-State Batteries:

Solid Electrolyte: Solid-state batteries use a solid electrolyte (instead of a liquid electrolyte in conventional batteries), which enhances safety by reducing the risk of leakage, thermal runaway, and fire hazards.

Higher Energy Density: Solid-state batteries have the potential to offer higher energy density compared to lithium-ion batteries, allowing for more energy storage capacity within the same volume.

Improved Performance: These batteries can potentially enable faster charging rates and higher power density, leading to reduced charging times and increased efficiency.

Advantages and Potential Benefits:

Enhanced Safety: Solid electrolytes are less prone to overheating and have lower flammability, improving overall battery safety.

Longer Lifespan: Solid-state battery technology has the potential to offer longer cycle life, reducing degradation and extending the battery's operational lifetime.

Environmental Impact: These batteries could potentially use more sustainable and environmentally friendly materials in their construction, reducing reliance on rare or hazardous elements.

Challenges and Current Status:

Technology Maturation: While promising, solid-state battery technology is still in the developmental stages, facing challenges related to manufacturing scalability and cost-effectiveness.

Material Compatibility: Finding suitable solid electrolyte materials that balance conductivity, stability, and manufacturability remains a critical challenge.

Research and Development:

Advancements in Materials: Ongoing research focuses on developing suitable solid electrolytes and electrode materials to improve performance and stability.

Manufacturing Innovation: Efforts are underway to optimize production processes and scale up manufacturing techniques for commercial viability.

Market and Future Prospects:

Commercialization Timeline: While not yet widely available, companies and research institutions are actively working toward commercializing solid-state batteries within the next decade.

Impact on Various Sectors: Solid-state batteries could revolutionize various industries, including electric vehicles (EVs), consumer electronics, and grid energy storage, by offering safer, higher-capacity, and more efficient energy storage solutions.

Solid-state batteries hold great promise in revolutionizing energy storage technology by potentially offering higher energy density, improved safety, and faster charging times. As ongoing research and development efforts continue to address technological challenges, solid-state batteries could play a pivotal role in advancing the efficiency and safety of energy storage systems, contributing to the proliferation of electric vehicles and enhancing energy storage across various applications.

Flow Batteries: Flow batteries, which store energy in liquid electrolytes, offer scalability and longer lifespans compared to traditional batteries. Research focuses on enhancing their energy density and reducing costs.

flow batteries present a promising energy storage solution characterized by their scalability, longevity, and potential for cost-effective large-scale deployment. Here are key aspects of flow batteries and ongoing research areas:

Flow Battery Technology:

Liquid Electrolytes: Flow batteries store energy in liquid electrolytes stored in external tanks, separating energy storage from power generation. The electrolytes flow through a cell stack to produce electricity.

Scalability: Their design allows for independent scaling of power and capacity by adjusting the size of electrolyte storage tanks, making them suitable for various applications, including grid-level energy storage.

Long Cycle Life: Flow batteries exhibit longer cycle life compared to conventional lithium-ion batteries due to the separation of energy and electrolyte storage, minimizing degradation.

Advantages and Features:

Energy Scalability: Flow batteries are well-suited for storing large amounts of energy over extended periods, making them ideal for renewable energy integration and grid-level applications.

Longevity: Their decoupled architecture enables long lifespans as the degradation of the cell components is slower than in traditional batteries.

Safety: The liquid electrolytes used in flow batteries are non-flammable and have low environmental impact, contributing to their safety.

Research Focus Areas:

Energy Density Improvement: Enhancing energy density is a key area of research to increase the amount of energy stored per unit volume or weight, making flow batteries more compact and efficient.

Cost Reduction: Researchers aim to reduce the cost of flow battery components, such as membranes, electrodes, and electrolytes, to make them more economically competitive.

Electrolyte Development: Innovations in electrolyte chemistry focus on improving performance, stability, and energy storage capacity of flow batteries.

Market and Deployment:

Grid-Level Storage: Flow batteries are particularly suitable for grid-level energy storage, supporting renewable energy integration and providing stability to the electrical grid.

Commercialization Efforts: Several companies are working on commercializing flow battery technology for various applications, including stationary energy storage and backup power systems.

Future Prospects:

Technological Advancements: Ongoing R&D efforts aim to overcome technical challenges, improve energy density, and reduce costs, making flow batteries more competitive in the energy storage market.

Increased Adoption: As advancements continue and costs decrease, flow batteries are expected to gain traction, especially in scenarios requiring large-scale, long-duration energy storage.

Flow batteries offer unique advantages in scalability, longevity, and safety, making them promising contenders for large-scale energy storage applications. Ongoing research focusing on improving energy density, reducing costs, and enhancing overall performance is expected to further unlock the potential of flow battery technology for supporting renewable energy integration and grid stability, paving the way for broader adoption in the future.

Thermal Energy Storage: Utilizing materials that store thermal energy, such as molten salt or phase-change materials, allows for storing excess energy as heat for later use in industrial processes or electricity generation.

Thermal energy storage (TES) systems are crucial for storing excess energy in the form of heat, offering solutions for various applications, including industrial processes and electricity generation. Here's an overview:

Types of Thermal Energy Storage:

Sensible Heat Storage: Involves storing heat by raising the temperature of a material without changing its phase, commonly using materials like rocks, water, or ceramics.

Latent Heat Storage: Stores energy by changing a material's phase, such as from solid to liquid or vice versa, utilizing phase-change materials (PCMs) like paraffin wax or molten salts.

Molten Salt Thermal Storage:

High Heat Capacity: Molten salts, such as a mixture of sodium nitrate and potassium nitrate, possess high heat capacities, making them efficient for storing and releasing thermal energy.

Applications: Often used in Concentrated Solar Power (CSP) plants, molten salt serves as a medium to store solar energy captured during the day for electricity generation during periods of lower sunlight or at night.

Phase-Change Materials (PCMs):

Energy Storage: PCMs absorb and release large amounts of latent heat during phase transitions, offering efficient energy storage in a small volume.

Applications: Used in buildings for passive temperature control, in electronics for thermal management, and in solar thermal applications for energy storage.

Advantages of Thermal Energy Storage:

Energy Time-Shifting: Allows excess energy generated during off-peak periods to be stored and used during peak demand times, balancing energy supply and demand.

Grid Stability: Offers grid stability by providing dispatchable energy, which helps manage fluctuations in renewable energy sources like solar or wind.

Challenges and Considerations:

Cost and Efficiency: The cost-effectiveness and overall efficiency of TES systems depend on the choice of materials and system design.

Material Stability: Ensuring material stability and reliability over multiple charging-discharging cycles is crucial for the longevity of thermal storage systems.

Future Developments:

Advanced Materials: Research focuses on developing new materials with enhanced heat storage capacity and improved thermal properties for more efficient TES.

Integration with Renewables: Increasing integration of TES systems with renewable energy sources like solar and wind power for grid stabilization and 24/7 clean energy supply.

Thermal energy storage using materials like molten salts and phase-change materials enables the efficient and flexible storage of excess heat or energy. These systems play a crucial role in supporting renewable energy integration, grid stability, and meeting energy demands during peak periods. Continued advancements in materials and system designs will further enhance the effectiveness and applicability of thermal energy storage across various industries, contributing to a more sustainable energy landscape.

Carbon Capture and Utilization (CCU):

Direct Air Capture (DAC): DAC technologies remove carbon dioxide directly from the atmosphere. Innovations aim to make this process more efficient and cost-effective, allowing captured CO_2 to be used in various applications or stored underground.

Carbon Utilization: Research focuses on utilizing captured CO_2 to produce valuable products, such as synthetic fuels, building materials, chemicals, or even to enhance agricultural processes.

Enhanced Carbon Sequestration: Techniques to enhance the natural process of carbon sequestration, such as reforestation, soil carbon sequestration, and ocean-based solutions, continue to evolve.

Green Hydrogen:

Electrolysis Technology: Advancements in electrolysis, particularly using renewable electricity, aim to produce green hydrogen by splitting water into hydrogen and oxygen. Innovations in electrolyzers aim to increase efficiency and reduce costs.

Advancements in electrolysis technology are pivotal in the production of green hydrogen, which involves the use of renewable electricity to split water molecules into hydrogen and oxygen. Here's an overview of electrolysis and its role in the production of green hydrogen:

Electrolysis for Green Hydrogen Production:

Process Overview: Electrolysis is the process of using an electrical current to split water (H_2O) into hydrogen (H_2) and oxygen (O_2) through electrochemical reactions.

Renewable Energy Integration: Green hydrogen production via electrolysis relies on renewable energy sources, such as solar or wind power, ensuring that the electricity used in the process is emission-free.

TYPES OF ELECTROLYZERS:

Proton Exchange Membrane (PEM) Electrolyzers: These operate at relatively low temperatures and are suitable for small to medium-scale applications, offering fast response times and high efficiency.

Alkaline Electrolyzers: These are more established and operate at higher temperatures. They are often used in larger-scale applications due to their lower cost but have slightly lower efficiency compared to PEM electrolyzers.

Advancements and Innovations:

Efficiency Improvements: Research aims to enhance the efficiency of electrolysis processes to minimize energy losses during hydrogen production.

Cost Reduction: Innovations focus on reducing the costs associated with electrolyzers, such as the use of cheaper materials or more efficient manufacturing processes.

Importance of Green Hydrogen:

Carbon Neutrality: Green hydrogen production avoids carbon emissions, contributing to a carbon-neutral or even carbon-negative energy cycle when produced using renewable energy sources.

Energy Storage and Sector Integration: Green hydrogen serves as a versatile energy carrier and can be used in various sectors, including transportation, industry, and energy storage, aiding in the integration of renewables.

Market Expansion and Deployment:

Growing Demand: Industries like transportation, heavy-duty vehicles, steel production, and chemical manufacturing show increasing interest in green hydrogen as a clean energy alternative.

Scaling Up Production: Efforts are underway to scale up green hydrogen production through large-scale electrolysis projects and collaborations between governments and private entities.

Future Prospects:

Technological Refinement: Continued R&D aims to improve electrolyzer efficiency, durability, and cost-effectiveness, making green hydrogen more competitive with other energy sources.

Policy Support: Favourable policies, incentives, and investments in renewable energy and hydrogen infrastructure are crucial for widespread adoption and market growth.

Advancements in electrolysis technology, especially utilizing renewable electricity, are instrumental in producing green hydrogen, offering a clean and versatile energy carrier. Continued innovation in electrolyzer technology and increased deployment of green hydrogen are essential steps toward achieving a sustainable and decarbonized energy future.

Hydrogen Storage: Research focuses on developing efficient and safe methods to store and transport hydrogen, such as solid-state hydrogen storage and chemical hydrogen carriers.

hydrogen storage is a critical aspect of utilizing hydrogen as an energy carrier, and ongoing research aims to develop efficient and safe methods for storing and transporting hydrogen. Here's an overview of some innovative approaches in hydrogen storage:

Solid-State Hydrogen Storage:

Metal Hydrides: Materials like complex metal hydrides can absorb and release hydrogen, providing a solid-state storage method. Research focuses on improving their hydrogen storage capacity, release kinetics, and operating temperatures.

Nanostructured Materials: Nanostructured materials, such as carbon nanotubes or metal-organic frameworks (MOFs), show potential for efficiently storing hydrogen due to their high surface area and tuneable properties.

Chemical Hydrogen Carriers:

Liquid Organic Hydrogen Carriers (LOHCs): These carriers chemically bond with hydrogen and release it upon demand. Research aims to enhance their hydrogen storage capacity, stability, and recyclability.

Ammonia (NH3): Ammonia is a promising hydrogen carrier due to its high hydrogen content by weight. Investigations focus on safe ammonia production, storage, and utilization in fuel cells or as a hydrogen source.

Advancements in Storage Technologies:

Efficiency Improvement: Research aims to enhance storage efficiency by optimizing materials and storage conditions to increase hydrogen capacity, improve kinetics, and reduce energy requirements for storage and release.

Safety Considerations: Addressing safety concerns related to hydrogen storage, such as preventing leaks, managing high pressures or temperatures, and ensuring compatibility with existing infrastructure.

Application in Transportation and Industry:

Fuel Cell Vehicles: Efficient hydrogen storage methods are crucial for fuel cell vehicles, ensuring adequate range and refuelling convenience comparable to conventional vehicles.

Industrial Applications: Reliable storage and transport methods are essential for industrial processes that utilize hydrogen as a feedstock or energy source, such as refining, chemical manufacturing, or power generation.

Future Prospects:

Material Innovation: Continued research focuses on developing novel materials and improving existing ones for efficient, safe, and cost-effective hydrogen storage.

System Integration: Integrating hydrogen storage methods with fuelling infrastructure and industrial applications requires a holistic approach for seamless adoption and utilization.

Hydrogen storage is a key aspect of utilizing hydrogen as an energy carrier in various sectors. Research efforts aim to develop innovative storage methods, such as solid-state storage and chemical carriers, that offer efficient, safe, and practical solutions for storing and transporting hydrogen. Advancements in storage technologies will play a crucial role in the widespread adoption of hydrogen as a clean and versatile energy source across industries.

Hydrogen Applications: Efforts are being made to expand the use of green hydrogen in sectors like transportation, industry, and energy production, aiming to replace fossil fuels and reduce emissions.

There is a growing push to expand the utilization of green hydrogen across various sectors as a clean and sustainable alternative to fossil fuels. Here's an overview of efforts aimed at deploying green hydrogen in key sectors:

Transportation:

Fuel Cell Vehicles (FCVs): Green hydrogen can power fuel cell vehicles, offering zero-emission transportation. Efforts focus on developing hydrogen refuelling infrastructure and advancing fuel cell technology for commercial vehicles, buses, and passenger cars.

Maritime and Aviation: Exploration of hydrogen as a potential fuel for ships and aircraft to decarbonize the transportation sector and reduce emissions from marine and aviation activities.

Industry:

Industrial Processes: Hydrogen serves as a clean feedstock in various industries, including refining, chemical production, metallurgy, and ammonia production, enabling cleaner processes and reducing carbon-intensive operations.

Power Generation: Green hydrogen can be used in gas turbines or fuel cells to generate electricity, providing a clean energy source for power plants, especially in locations where renewable electricity may not be consistently available.

Energy Production and Storage:

Grid Balancing: Hydrogen can serve as an energy storage medium, providing grid stability by storing excess renewable energy and supplying it during high-demand periods, supporting the integration of renewables.

Heat Generation: Hydrogen combustion or utilization in fuel cells can provide heat for industrial processes or district heating, reducing reliance on fossil fuels for heat generation.

Challenges and Opportunities:

Infrastructure Development: Expanding hydrogen infrastructure, including production facilities, storage, transportation, and refuelling stations, is crucial for widespread adoption.

Cost Competitiveness: Continued efforts aim to reduce the cost of green hydrogen production through advancements in electrolysis technology, renewable energy, and economies of scale.

Policy Support and Initiatives:

Government Incentives: Supportive policies, subsidies, and incentives aim to promote the adoption of green hydrogen across sectors and drive investment in hydrogen technologies.

International Collaboration: Collaborative efforts among countries, industry players, and research institutions aim to accelerate the development and deployment of green hydrogen on a global scale.

Future Outlook:

Market Expansion: The growing interest in decarbonization and the transition to cleaner energy sources are expected to drive increased demand and investment in green hydrogen technologies.

Technological Advancements: Continuous innovation and R&D will lead to improved efficiency, scalability, and cost-effectiveness, making green hydrogen a more viable and competitive energy option.

The widespread adoption of green hydrogen across transportation, industry, and energy production holds the potential to significantly reduce carbon emissions and drive the transition towards a more sustainable and decarbonized future. Efforts in infrastructure development, technological advancements, and supportive policies are crucial in realizing the full potential of green hydrogen as a clean and versatile energy carrier across various sectors.

Other Emerging Clean Technologies:

Advanced Nuclear Reactors: Next-generation nuclear reactor designs focus on increased safety, reduced waste, and enhanced efficiency to provide a low-carbon energy source.

advancements in next-generation nuclear reactor designs aim to address various challenges associated with traditional nuclear reactors while offering enhanced safety, reduced waste, and improved efficiency as a low-carbon energy source. Here's an overview:

Safety Innovations:

Passive Safety Features: Advanced reactors incorporate passive safety systems that operate without human intervention or external power sources, enhancing safety during unforeseen events.

Improved Cooling Systems: Innovative cooling systems prevent overheating and mitigate potential accidents, ensuring better heat removal and reactor stability.

Reduced Waste and Fuel Utilization:

Advanced Fuel Cycles: Some reactor designs aim to utilize different fuel cycles, such as fast reactors or molten salt reactors, to extract more energy from nuclear fuel and reduce the volume and longevity of nuclear waste.

Transmutation Technologies: Certain reactor designs explore transmutation processes to convert long-lived nuclear waste into shorter-lived isotopes or usable fuel, minimizing the environmental impact of nuclear waste.

Efficiency and Cost-Effectiveness:

Higher Efficiency: Next-gen reactors aim to improve thermal efficiency and power output, maximizing energy generation from nuclear fuel while reducing operational costs.

Modular Designs: Some designs focus on modularity, enabling smaller, scalable reactors that could potentially be constructed more quickly and economically compared to traditional large-scale reactors.

Novel Reactor Technologies:

Small Modular Reactors (SMRs): SMRs offer scalable, more flexible options for nuclear power generation, suitable for diverse applications, including remote locations or powering industrial facilities.

Molten Salt Reactors (MSRs): MSRs use liquid fuel, offering improved safety features, enhanced fuel utilization, and potential applications in electricity generation and process heat production.

Regulatory and Deployment Challenges:

Regulatory Approval: Developing new designs requires navigating regulatory hurdles, ensuring compliance with safety standards and licensing procedures.

Public Perception: Addressing public concerns about nuclear energy safety, waste management, and proliferation risks remains a challenge for wider acceptance and deployment.

Research and Development:

Public-Private Partnerships: Collaboration between governments, research institutions, and private entities supports R&D efforts, enabling technological advancements and deployment of advanced reactor designs.

Innovative Materials: Advancements in materials science aim to develop radiation-resistant materials for reactor components, ensuring longevity and safety.

Future Outlook:

Carbon-Free Energy: Advanced nuclear reactors, with their potential for increased safety, reduced waste, and improved efficiency, could play a significant role in providing reliable, low-carbon energy to help combat climate change.

Deployment Challenges: While promising, the widespread deployment of next-generation reactors may require overcoming technical, regulatory, and societal challenges, necessitating sustained investment and public acceptance.

Next-generation nuclear reactor designs offer promising solutions to the challenges associated with traditional reactors, aiming for enhanced safety, reduced waste, and improved efficiency in providing low-carbon energy. Continued R&D, regulatory support, and public engagement will be crucial in realizing the full potential of advanced nuclear technologies as a sustainable and reliable energy source.

Smart Grid Technology: Innovations in grid management, including AI-driven demand response systems and advanced monitoring, improve grid stability, reliability, and integration of renewable energy sources.

Advancements in smart grid technology are transforming traditional electricity grids into more efficient, resilient, and flexible systems. Here's an overview of innovations in smart grid management:

Demand Response Systems:

AI-Driven Demand Response: Artificial Intelligence (AI) algorithms analyse data to predict and optimize energy demand, enabling utilities to adjust consumption patterns, reduce peak demand, and balance loads more efficiently.

Smart Appliances and Meters: Integration of smart appliances and meters allows consumers to participate in demand response programs, enabling them to adjust energy usage based on real-time pricing or grid signals.

Advanced Monitoring and Control:

Sensors and IoT Integration: Deployment of sensors and Internet of Things (IoT) devices across the grid infrastructure facilitates real-time monitoring of electricity flows, grid health, and equipment performance.

Predictive Maintenance: AI-based analytics and machine learning help predict and prevent potential equipment failures, reducing downtime and improving grid reliability.

Grid Integration of Renewables:

Grid-Friendly Renewables: Advanced grid management systems enable the smooth integration of variable renewable energy sources like solar and wind by forecasting generation and adjusting grid operations accordingly.

Virtual Power Plants (VPPs): Aggregation of distributed energy resources (DERs), including rooftop solar panels and battery storage, into VPPs enhances grid stability and enables better management of fluctuating renewable energy outputs.

Grid Resilience and Flexibility:

Microgrids: Implementation of microgrids allows for localized energy generation and distribution, enhancing resilience during grid outages and natural disasters.

Energy Storage Integration: Smart grid technologies facilitate the integration of energy storage systems, enabling the storage of excess renewable energy for use during peak demand or grid disruptions.

Data Analytics and Decision Support:

Big Data Analytics: Utilizing large datasets and analytics helps grid operators make informed decisions, optimize grid operations, and plan infrastructure upgrades more effectively.

AI-Based Grid Optimization: AI algorithms optimize grid configurations, improve energy flow, and enhance grid stability by dynamically adjusting operations in response to changing conditions.

Regulatory Support and Implementation Challenges:

Regulatory Frameworks: Supportive policies and regulations are crucial to incentivize investment in smart grid technologies and ensure interoperability among diverse grid systems.

Deployment Challenges: Challenges include legacy infrastructure, cybersecurity concerns, and the need for skilled professionals to manage and maintain advanced grid technologies.

FUTURE PROSPECTS:

Energy Transition: Smart grid advancements are pivotal in facilitating the transition to a more decentralized, cleaner, and resilient energy system, accommodating increased renewable energy integration.

Technological Innovation: Continued innovation in smart grid technologies, coupled with supportive policies, will further enhance grid flexibility, efficiency, and reliability, benefiting both utilities and consumers.

Smart grid technologies, driven by AI, IoT, and advanced analytics, play a crucial role in modernizing electricity grids, enabling efficient management, integrating renewable energy sources, and enhancing grid reliability and resilience. Continued innovation and strategic deployment of these technologies are pivotal for a sustainable, reliable, and resilient energy future.

Circular Economy Solutions: Technologies promoting a circular economy, such as advanced recycling methods, sustainable materials, and waste-to-energy processes, aim to minimize resource consumption and waste generation. Continued research and development in these areas are likely to yield further advancements and drive the adoption of cleaner and more sustainable technologies. For the most current information, I recommend consulting recent publications, academic research, and updates from industry leaders and clean technology organizations.

Air Quality Monitoring

AI (Artificial Intelligence), big data analytics, and IoT (Internet of Things) play pivotal roles in revolutionizing emissions monitoring, management, and optimization in various industries. Here's a breakdown of their contributions:

Emissions Monitoring:

Sensor Technology and IoT: IoT devices equipped with sensors are deployed in industries and infrastructure to collect real-time data on emissions, air quality, and environmental parameters. These sensors continuously monitor various pollutants, allowing for immediate detection and response to anomalies.

The deployment of Internet of Things (IoT) devices equipped with sensors has revolutionized environmental monitoring across industries and infrastructure. These sensors play a vital role in collecting real-time data on various environmental parameters, emissions, and air quality, enabling immediate detection and response to anomalies. Here's an overview:

Environmental Monitoring:

Air Quality Sensors: These sensors measure pollutants like particulate matter (PM), nitrogen oxides (NOx), sulfur dioxide (SO2), volatile organic compounds (VOCs), and carbon monoxide (CO) in the air.

Emissions Monitoring: Sensors track emissions from industrial processes, power plants, transportation, and other sources, helping to ensure compliance with environmental regulations and identify potential issues.

Real-Time Data Collection:

Continuous Monitoring: IoT-enabled sensors provide continuous, real-time data collection, allowing for the immediate detection of variations in environmental parameters.

Remote Sensing: Sensors deployed across large geographical areas or remote locations provide valuable data without the need for physical presence, enabling efficient monitoring and timely responses.

Anomaly Detection and Response:

Early Warning Systems: IoT sensors equipped with AI algorithms can detect abnormal pollutant levels or environmental changes, triggering alerts for swift corrective actions.

Predictive Analytics: Data collected by sensors can be analysed to predict potential environmental issues, enabling proactive measures to mitigate risks or prevent incidents.

Applications Across Industries:

Industrial Sector: Monitoring emissions and air quality in factories, refineries, and manufacturing plants to ensure compliance with environmental standards and optimize processes.

Urban Environments: Sensors deployed in cities monitor air quality, helping local governments implement measures to improve public health and reduce pollution.

Benefits and Challenges:

Benefits: Real-time data collection and analysis enable prompt responses, better regulatory compliance, improved public health, and more efficient resource utilization.

Challenges: Issues include sensor accuracy, calibration, data privacy, and the need for standardized protocols for data collection and interpretation.

Future Outlook:

Advancements in Sensor Technology: Continued innovation will lead to more accurate, affordable, and durable sensors capable of measuring a broader range of pollutants.

Integration with Smart Systems: Integration of sensor data with smart city initiatives, AI-driven analytics, and decision support systems will enhance environmental management and policymaking.

IoT devices equipped with sensors are transforming environmental monitoring by providing real-time data on emissions, air quality, and environmental parameters. Continuous monitoring and immediate response capabilities enable industries and governments to make informed decisions, mitigate environmental risks, and work towards a healthier and more sustainable future.

Data Collection and Integration: These technologies enable the collection, aggregation, and integration of vast amounts of data from multiple sources, including sensors, satellites, and other monitoring systems, providing a comprehensive view of emissions across different sectors.

Advancements in technology have facilitated the collection, aggregation, and integration of extensive datasets from diverse sources, enabling a comprehensive view of emissions across various sectors. Here's an overview:

Multi-Source Data Collection:

Sensor Networks: IoT sensors placed in industries, urban areas, and infrastructure continuously collect real-time data on emissions, air quality, and environmental parameters.

Satellite Monitoring: Satellite-based sensors provide a broader geographical perspective, capturing emissions, changes in land use, deforestation, and other environmental indicators.

Monitoring Systems: Specialized monitoring systems installed in power plants, refineries, and transportation hubs track emissions directly from these sources.

Data Aggregation and Integration:

Big Data Platforms: Advanced data management platforms process and aggregate information from various sources, ensuring compatibility and harmonization of datasets.

Cloud Computing: Cloud-based solutions facilitate the storage, management, and analysis of vast datasets, allowing for scalable and efficient data processing.

GIS and Spatial Analysis: Geographic Information Systems (GIS) integrate spatial data, enabling visual representations and spatial analysis of emissions and their sources.

Comprehensive Emissions Mapping:

Sectoral Analysis: Integration of data from multiple sources allows for a comprehensive breakdown of emissions by sectors such as industry, transportation, energy, and agriculture.

Temporal Analysis: Continuous data collection enables temporal analysis, identifying emission patterns over time, daily or seasonal variations, and long-term trends.

Benefits of Integrated Data:

Holistic View: Integration of diverse data sources provides a holistic understanding of emissions, aiding in policy formulation, urban planning, and environmental management.

Improved Decision-Making: Comprehensive data assists stakeholders in making informed decisions, targeting emission reduction strategies, and prioritizing interventions.

Challenges and Considerations:

Data Quality and Standardization: Ensuring data accuracy, consistency, and standardization across different sources and formats is essential for reliable analysis.

Privacy and Security: Protecting sensitive data and ensuring privacy while sharing information across different entities remain significant concerns.

Future Directions:

AI and Predictive Analytics: Utilizing AI and machine learning for predictive analytics enhances the capability to forecast emissions, enabling proactive interventions.

Policy and Governance: Integrating data insights into policy frameworks and governance structures can drive more effective emission reduction strategies.

DATA COLLECTION AND integration from various sources, including sensors, satellites, and monitoring systems, offer a comprehensive view of emissions across sectors and geographies. This integrated data serves as a critical resource for policymakers, urban planners, and environmental agencies, facilitating evidence-based decision-making and targeted actions to mitigate emissions and address environmental challenges.

AI and Machine Learning for Analysis: AI algorithms process the collected data to identify patterns, trends, and potential issues related to emissions. Machine learning models can detect anomalies, predict emissions levels, and provide insights into optimizing processes to reduce emissions.

AI and machine learning play pivotal roles in analysing the vast amounts of collected data related to emissions, offering insights, predictions, and optimization strategies. Here's an overview of their applications in emissions analysis:

Pattern Recognition and Trend Identification:

Pattern Detection: AI algorithms analyse historical emissions data to identify recurring patterns, such as seasonal variations or trends associated with specific activities or industries.

Trend Identification: Machine learning models detect long-term trends in emissions, helping stakeholders understand evolving patterns and drivers behind changes in emission levels.

Anomaly Detection and Prediction:

Anomaly Detection: AI-powered systems detect irregularities or anomalies in emissions data, signalling potential malfunctions, leaks, or deviations from expected norms that require attention.

Emissions Prediction: Machine learning models utilize historical data and contextual information to forecast future emissions levels, aiding in proactive planning and interventions.

Process Optimization for Emission Reduction:

Optimization Algorithms: AI-driven optimization models recommend strategies to minimize emissions by optimizing industrial processes, energy usage, or transportation routes.

Emission Reduction Insights: Machine learning identifies areas for emission reduction, offering insights into operational changes or technological advancements to achieve sustainability goals.

Adaptive Systems and Continuous Improvement:

Adaptive Control Systems: AI-based control systems dynamically adjust operations in real-time to optimize processes for reduced emissions while maintaining efficiency.

Continuous Learning: Machine learning models continuously improve accuracy and efficiency by learning from new emissions data, refining predictions, and recommendations.

Integration with Smart Systems:

Integrated Decision Support: AI integrates with decision support systems, providing actionable insights to policymakers, industries, and urban planners for emission mitigation strategies.

Smart Grids and Cities: AI applications assist in optimizing energy usage, traffic management, and infrastructure planning, contributing to lower emissions in smart cities.

Challenges and Future Developments:

Data Quality and Interpretability: Ensuring reliable and interpretable AI models requires high-quality, diverse, and standardized emissions data.

Ethical Considerations: Ethical usage of AI in emissions analysis involves transparency, fairness, and accountability in decision-making processes.

AI and machine learning empower emissions analysis by detecting patterns, predicting future levels, and suggesting strategies for emissions reduction. As these technologies evolve, their integration with decision-making processes and their capacity to drive proactive emission reduction strategies are expected to play a vital role in addressing environmental challenges and fostering sustainability.

Emissions Management and Forecasting

P redictive Analytics: AI-powered predictive analytics help in forecasting emissions based on historical data, weather patterns, production schedules, and other factors. This allows proactive measures to be taken to prevent or minimize emissions spikes.

AI-powered predictive analytics are instrumental in emissions management by leveraging historical data, weather patterns, production schedules, and other relevant factors to forecast emissions. This proactive approach aids in preventing or mitigating emissions spikes and allows for better emissions management. Here's an in-depth look at how predictive analytics assist in this aspect:

Forecasting Emissions:

Historical Data Analysis: AI algorithms analyse historical emissions data to identify patterns and trends, enabling the creation of predictive models.

Historical data analysis forms a crucial part of leveraging AI algorithms to identify patterns and trends, facilitating the creation of predictive models for emissions management. Here's an in-depth look at how historical data analysis contributes to this process:

Leveraging Historical Emissions Data:

Data Collection and Cleaning: Gathering comprehensive historical emissions data from various sources and ensuring its quality and consistency for analysis.

Pattern Identification: Utilizing AI algorithms to sift through large datasets, recognizing patterns, correlations, and trends in emissions data over time.

Identifying Emission Patterns and Trends:

Seasonal Variations: Analysing data to detect seasonal fluctuations or patterns in emissions due to weather, production cycles, or specific operational activities.

Event-Based Analysis: Examining historical data to identify emission spikes or anomalies associated with particular events, operations, or environmental conditions.

Model Development and Training:

Machine Learning Algorithms: Employing machine learning techniques such as regression, clustering, or neural networks to build predictive models based on historical emissions data.

Feature Selection: Determining which emission-related variables or factors are most influential in predicting future emission levels, refining model inputs.

Validation and Refinement:

Model Validation: Testing the predictive models against historical data not used during training to assess their accuracy, reliability, and ability to forecast emissions.

Continuous Improvement: Iteratively refining models by incorporating new data and insights to enhance accuracy and adaptability to changing emission patterns.

Predictive Modelling for Emissions Forecasting:

Forecasting Future Emissions: Utilizing the trained predictive models to forecast future emission levels based on historical trends and identified patterns.

Scenario Analysis: Assessing various scenarios by adjusting model inputs to predict emissions under different conditions or policy changes.

Benefits of Historical Data Analysis:

Informed Decision-Making: Insights from historical data aid in making informed decisions for emissions reduction strategies and operational improvements.

Proactive Management: Predictive models derived from historical data enable proactive measures to mitigate potential emissions spikes.

Challenges and Considerations:

Data Quality and Availability: Ensuring data accuracy, consistency, and accessibility across different sources is crucial for reliable historical analysis.

Interpretability and Transparency: Understanding the reasons behind identified patterns and ensuring the transparency of AI-driven models is important for trust and adoption.

Historical data analysis, powered by AI algorithms, plays a pivotal role in emissions management by identifying patterns, trends, and correlations in emissions data. The insights derived from historical analysis inform the development of predictive models, enabling proactive and data-driven strategies for emissions reduction and environmental sustainability.

Weather and Environmental Factors: Incorporating weather forecasts, seasonal variations, and other environmental parameters improves the accuracy of emissions predictions.

Integrating weather forecasts, seasonal variations, and other environmental factors significantly enhances the accuracy and reliability of emissions predictions. Here's an in-depth look at how these factors contribute to more precise emission forecasting:

Weather Forecast Integration:

Impact on Emissions: Weather elements like temperature, humidity, wind speed, and precipitation directly influence emission levels from various sources like transportation, industry, and agriculture.

Dynamic Emission Models: AI-powered models incorporating real-time or forecasted weather data dynamically adjust emission predictions, accounting for weather-related fluctuations.

Seasonal Variations:

Temperature Variability: Cold or hot seasons affect heating or cooling demands, impacting energy consumption and subsequently emissions from power plants or residential sources.

Agricultural Impact: Seasonal changes affect farming practices, leading to variations in emissions from agricultural activities like field burning or livestock management.

Environmental Parameters:

Air Quality Index (AQI): Integration of AQI data helps correlate emissions with local air quality, identifying regions prone to high pollution levels under specific conditions.

Topographical Influence: Terrain, urban layout, and proximity to natural barriers influence air circulation and pollutant dispersion, affecting localized emissions patterns.

Emission Sources Sensitivity to Weather:

Transportation Sector: Weather conditions impact vehicle efficiency and traffic flow, affecting emissions due to idling, congestion, or fuel consumption changes.

Power Generation: Weather affects renewable energy production (solar, wind) and influences demand for heating or cooling, altering power plant operations and emissions.

Benefits of Weather and Environmental Integration:

Enhanced Accuracy: Incorporating weather and environmental parameters refines predictive models, improving their ability to forecast emission variations.

Precise Mitigation Strategies: Understanding weather-related emission patterns aids in formulating targeted strategies to mitigate emissions during specific conditions.

Challenges and Considerations:

Data Integration: Ensuring seamless integration of diverse datasets (emissions, weather, environmental) for accurate analysis poses technical and compatibility challenges.

Model Complexity: Advanced modelling to incorporate numerous variables requires sophisticated algorithms and computing resources.

Future Developments:

Advanced Modelling Techniques: Continued advancements in AI and machine learning enable more sophisticated models that better account for intricate weather-emission correlations.

Real-Time Data Integration: Integration of real-time weather data into predictive models allows for more immediate and accurate emission forecasts.

Incorporating weather forecasts, seasonal variations, and environmental parameters significantly improves the accuracy of emissions predictions. Understanding the intricate relationship between weather conditions and emissions empowers predictive models, aiding in more precise forecasting and the formulation of targeted emission reduction strategies.

Proactive Measures:

Early Warning Systems: Predictive analytics provide early warnings about potential emissions spikes or deviations from expected levels, enabling proactive responses.

early warning systems powered by predictive analytics play a crucial role in emissions management by providing advance notice about potential spikes or deviations in emissions levels. These systems enable proactive responses, allowing for timely interventions to mitigate or prevent adverse environmental impacts. Here's an in-depth look at early warning systems in emissions management:

Proactive Monitoring:

Continuous Monitoring: Predictive models constantly analyse incoming data to detect patterns and deviations from expected emission levels in real-time.

Threshold Identification: Establishing thresholds or benchmarks enables the early detection of emissions exceeding predefined levels, signalling potential issues.

Early Detection of Anomalies:

Anomaly Identification: Predictive analytics swiftly identify irregularities or unexpected trends in emissions data that might indicate upcoming spikes or deviations.

Pattern Recognition: Machine learning algorithms recognize unusual patterns or sudden changes in emissions data, triggering alerts for further investigation.

Prompt Alert Systems:

Automated Alerts: Early warning systems trigger automated alerts or notifications when emissions reach or are predicted to exceed predetermined thresholds.

Notification Mechanisms: Alerts are transmitted to relevant stakeholders or responsible personnel, facilitating prompt action and intervention.

Proactive Response Strategies:

Preventive Measures: Advanced notice allows for immediate action, such as adjusting operations, implementing control measures, or conducting maintenance to avert emissions spikes.

Optimized Operations: Proactive responses enable industries to optimize processes or equipment, minimizing emissions while maintaining operational efficiency.

Continuous Improvement:

Model Refinement: Continuous evaluation and refinement of predictive models based on observed emissions data ensure the system's accuracy and effectiveness.

Feedback Loop: Incorporating feedback from responses and outcomes improves the predictive capability of the system for future emissions forecasts.

Benefits of Early Warning Systems:

Risk Mitigation: Timely alerts facilitate the mitigation of potential environmental risks or regulatory non-compliance associated with excessive emissions.

Resource Optimization: Proactive responses help allocate resources effectively, reducing the impact of emissions spikes on operations or the environment.

Challenges and Considerations:

Data Accuracy: Early warning systems rely on accurate and real-time data for reliable predictions, posing challenges in data quality and timeliness.

Response Time: Ensuring swift and effective responses to alerts requires streamlined communication and decision-making processes.

Future Directions:

Integration with Decision Support: Enhanced integration of early warning systems with decision support tools enables more informed and effective responses.

Advanced AI Techniques: Continued advancements in AI algorithms refine early warning systems, improving their predictive accuracy and efficiency.

Early warning systems powered by predictive analytics offer proactive monitoring and timely alerts about potential emissions spikes or deviations. These systems enable industries to take swift and targeted actions, minimizing environmental impacts, ensuring compliance, and optimizing operational efficiency in emissions management.

Optimized Operations: Anticipating emissions fluctuations allows industries to adjust production schedules or processes to minimize environmental impact.

Anticipating emissions fluctuations through predictive analytics empowers industries to adjust their production schedules, operational processes, and resource allocation strategies to minimize environmental impact. Here's a deeper dive into how optimized operations can mitigate environmental impact:

Emissions-Aware Operations:

Dynamic Planning: Predictive analytics forecasting emission spikes or fluctuations prompts industries to plan and adapt operations accordingly.

Real-time Adjustments: Monitoring emissions trends in real-time allows for immediate adjustments to minimize environmental impact.

Production Schedule Adjustments:

Efficient Resource Use: Adjusting production schedules in anticipation of high-emission periods optimizes resource utilization and energy consumption.

Peak Demand Management: Aligning production schedules with lower emission periods helps manage energy demand during peak times.

Process Optimization:

Technology Adaptation: Switching to cleaner or more efficient technologies during periods of high emissions helps reduce overall environmental impact.

Efficiency Improvements: Optimizing manufacturing processes or equipment to reduce emissions while maintaining productivity levels.

Emission Reduction Strategies:

Emission Controls: Implementing emission control technologies or practices during anticipated high-emission periods to mitigate environmental impact.

Alternate Raw Materials: Using alternative materials or substitutes that have lower emission footprints for production processes.

Environmental Impact Mitigation:

Compliance Measures: Proactive adjustments aid industries in adhering to emission limits and environmental regulations.

Stakeholder Responsibility: Demonstrating responsibility towards environmental concerns by mitigating emissions fluctuations.

Benefits of Optimized Operations:

Environmental Sustainability: Minimizing emissions fluctuations reduces the overall environmental footprint and contributes to sustainability goals.

Cost Savings: Optimized operations often lead to energy and resource savings, reducing operational costs in the long run.

Challenges and Considerations:

Balancing Efficiency and Impact: Striking a balance between maintaining operational efficiency while minimizing emissions remains a challenge.

Technological Adaptation: Adapting to newer, cleaner technologies or practices might require investment and transition periods.

Future Developments:

Smart Automation: Increased automation and AI-driven systems enable faster and more precise adjustments to operational schedules.

Renewable Integration: Incorporating renewable energy sources into production processes for more sustainable operations.

Anticipating emissions fluctuations through predictive analytics empowers industries to adjust production schedules, optimize processes, and adopt emission reduction strategies. These proactive measures not only minimize environmental impact but also contribute to cost savings and sustainability, aligning with global efforts towards a greener future.

Factors Considered in Predictive Modelling:

Production Schedules: Understanding production cycles and schedules aids in predicting emissions during high-output periods or specific manufacturing processes.

Comprehending production cycles and schedules is instrumental in predicting emissions, particularly during high-output periods or specific manufacturing processes. Here's a closer look at how understanding production schedules assists in emissions prediction:

Emissions Prediction and Production Cycles:

Routine Emissions Patterns: Historical data analysis of emissions during different production phases or cycles helps identify recurring patterns.

High-Output Periods: Anticipating increased emissions during peak production times or specific manufacturing phases due to heightened activity.

Factors Impacting Emissions:

Volume-Driven Emissions: More significant output typically correlates with increased energy consumption and emissions from machinery or processes.

Process-Specific Emissions: Certain manufacturing stages or processes might inherently produce higher emissions due to the nature of operations involved.

Operational Insights:

Production Schedule Analysis: Examining production schedules provides insights into periods when machinery or facilities are in high-demand, correlating with elevated emissions.

Resource Utilization: Identifying resource-intensive phases helps predict emissions linked to the usage of energy, raw materials, or specific chemicals.

Tailored Emission Reduction Strategies:

Focused Mitigation: Knowing emission hotspots within production cycles aids in tailoring targeted mitigation strategies for those specific phases.

Process Optimization: Introducing cleaner technologies or improving efficiency during high-emission production stages to minimize environmental impact.

Continuous Improvement:

Model Refinement: Continuous evaluation of predictive models based on observed emission patterns during production cycles ensures accuracy and relevance.

Adaptive Strategies: Incorporating feedback from emissions data during various production cycles enhances the efficacy of emission reduction plans.

Benefits of Understanding Production Schedules:

Proactive Planning: Enables proactive measures to mitigate emissions during high-output periods, avoiding last-minute reactive strategies.

Resource Allocation: Facilitates efficient resource allocation and investment in emission reduction technologies during critical production phases.

Challenges and Considerations:

Data Availability: Access to accurate and detailed production schedules is crucial for reliable emissions prediction.

Complexity of Operations: Diverse manufacturing processes may require tailored emission reduction strategies, adding complexity to planning.

Future Directions:

Integrated Systems: Improved integration of production planning systems with emission prediction models for more accurate forecasting.

Industry Collaboration: Sharing best practices and emission data among industries to collectively develop effective emission reduction strategies.

Understanding production cycles and schedules aids in predicting emissions during high-output periods or specific manufacturing processes. This knowledge allows industries to foresee emission trends, enabling targeted strategies to reduce environmental impact and optimize resources efficiently. It forms a critical aspect of proactive emission management within industrial operations.

External Influences: External factors like temperature, humidity, wind patterns, and their impact on emissions are considered for more accurate predictions.

considering external influences such as temperature, humidity, wind patterns, and other environmental factors is crucial for accurate emissions predictions. These external variables significantly affect emissions from various sources, and incorporating them into predictive models enhances the precision of emission forecasts. Here's a deeper insight into their impact:

Temperature and Weather Conditions:

Energy Demand: Temperature variations influence heating and cooling needs, impacting energy consumption and subsequent emissions from heating systems or air conditioning.

Chemical Reactions: Certain emissions, especially from combustion processes, are temperature-sensitive and can vary based on temperature changes.

Humidity and Moisture Levels:

Combustion Efficiency: Humidity affects combustion efficiency, potentially altering emissions levels from combustion-based machinery or systems.

Chemical Processes: Moisture levels can influence chemical reactions in industrial processes, impacting emission rates.

Wind Patterns and Air Circulation:

Dispersion of Pollutants: Wind direction and speed determine the dispersion of emitted pollutants, impacting the concentration and distribution of emissions in the atmosphere.

Localized Emissions: Wind patterns may affect emissions' local impact, especially in areas with specific topographies or where emissions from neighbouring sources converge.

Impact on Different Industries:

Transportation: Wind patterns influence vehicle aerodynamics and fuel efficiency, affecting emissions from the transportation sector.

Power Generation: Wind speeds can influence the efficiency of wind turbines, impacting electricity generation and associated emissions.

Predictive Modelling Considerations:

Integration with Models: Incorporating weather and environmental data into predictive models refines emissions predictions, accounting for external influences' effects.

Dynamic Adjustments: Real-time weather data allows for dynamic adjustments in emission forecasts based on changing environmental conditions.

Benefits of Considering External Influences:

Accurate Predictions: Incorporating external factors leads to more precise emission forecasts, aiding in proactive emission management strategies.

Targeted Interventions: Understanding how weather impacts emissions enables targeted interventions during specific weather-related emission spikes.

Challenges and Considerations:

Data Integration: Obtaining and integrating real-time weather data with emissions data for accurate predictions poses technical challenges.

Model Complexity: Accounting for multiple external variables adds complexity to predictive models, requiring sophisticated algorithms and computing resources.

Future Developments:

Advancements in AI Modelling: AI-driven models that efficiently integrate and process diverse environmental data for more accurate predictions.

Localized Predictions: Improved capability to predict localized emissions' impact under varying environmental conditions for precise interventions.

Incorporating external influences like temperature, humidity, wind patterns, and other environmental factors into emissions predictive models is crucial. Understanding their impact allows industries to generate more accurate forecasts, enabling proactive measures and targeted interventions to manage and mitigate emissions more effectively across different sectors.

Real-Time Adjustments:

Dynamic Decision-Making: Predictive models enable real-time adjustments, allowing industries to optimize operations based on forecasted emission levels.

Predictive models empower industries to make real-time adjustments and optimize operations based on forecasted emission levels. This dynamic decision-making process plays a vital role in minimizing emissions and optimizing resource utilization. Here's a closer look at how predictive models facilitate real-time adjustments for emission management:

Real-Time Decision-Making:

Continuous Monitoring: Predictive models continuously analyse incoming data, providing updated forecasts and insights into future emission trends.

Immediate Responses: Real-time emission forecasts trigger prompt responses to optimize operations for minimal environmental impact.

Optimization Strategies:

Operational Adjustments: Adjusting production rates, machinery utilization, or process parameters in response to forecasted emission levels to maintain efficiency while reducing emissions.

Resource Allocation: Allocating resources efficiently based on emission forecasts, ensuring optimal utilization while minimizing environmental impact.

Adaptive Control Systems:

Automated Control: AI-driven systems can automatically adjust operations or machinery settings based on real-time emission predictions, optimizing performance.

Smart Technologies: Implementation of IoT devices and sensors aids in collecting real-time data, facilitating immediate adjustments in response to emission forecasts.

Proactive Mitigation:

Preventive Measures: Proactively implementing emission reduction measures or switching to cleaner technologies based on predicted emission levels.

Dynamic Planning: Altering schedules, workflows, or maintenance activities to align with anticipated emission fluctuations.

Benefits of Real-Time Adjustments:

Emission Reduction: Immediate responses to forecasted emission levels help in reducing overall emissions during high-risk periods.

Operational Efficiency: Optimizing operations based on real-time predictions maintains productivity while reducing environmental impact.

Challenges and Considerations:

Technological Integration: Ensuring seamless integration of predictive models with operational systems for timely adjustments.

Accuracy of Predictions: Reliability of real-time predictions depends on the accuracy and precision of the underlying predictive models.

Future Directions:

Enhanced Automation: Further automation and AI-driven systems for quicker and more accurate real-time adjustments.

Advanced Sensor Networks: Improved sensor technologies and data collection systems for more precise real-time monitoring and forecasting.

Real-time adjustments based on forecasted emission levels, facilitated by predictive models, allow industries to optimize operations for reduced emissions without compromising productivity. This dynamic decision-making approach enables proactive emission management, contributing to environmental sustainability while maintaining operational efficiency.

Resource Allocation: Predictive analytics assist in resource allocation, enabling the allocation of mitigation resources to areas predicted to have high emissions.

Predictive analytics play a pivotal role in resource allocation by identifying and directing mitigation resources to areas predicted to have high emissions. This strategic allocation based on predictive models enables efficient and targeted mitigation efforts. Here's a closer look at how predictive analytics assist in resource allocation for emissions management:

Predictive Models for Resource Allocation:

Emission Hotspots Identification: Predictive models highlight areas or processes likely to produce high emissions based on historical data and current trends.

Scenario Planning: Predictive analytics enable the creation of scenarios based on emission forecasts, aiding in proactive resource allocation strategies.

Targeted Mitigation Efforts:

Focused Interventions: Concentrating mitigation resources like emission control technologies, maintenance, or process modifications in areas identified as high-risk for emissions.

Process Optimization: Allocating resources towards optimizing specific operations or technologies prone to higher emissions, reducing their environmental impact.

Efficient Deployment:

Cost-Effective Allocation: Targeted allocation of mitigation resources optimizes their use, avoiding unnecessary expenses by focusing efforts where they're most needed.

Prioritization of Efforts: Identifying critical areas for emission reduction allows for prioritizing mitigation efforts based on their potential environmental impact.

Flexibility in Resource Utilization:

Adaptive Planning: Predictive models enable flexible resource allocation, allowing adjustments as emission forecasts evolve or operational conditions change.

Dynamic Allocation: Shifting resources in real-time based on updated emission forecasts for immediate and effective mitigation strategies.

Benefits of Predictive Resource Allocation:

Emission Reduction: Targeted allocation minimizes emissions in critical areas, contributing to overall environmental improvement.

Optimal Resource Utilization: Efficient allocation ensures the best use of mitigation resources, enhancing their effectiveness in reducing emissions.

Challenges and Considerations:

Data Accuracy: Reliability of predictive models heavily relies on the accuracy and completeness of emissions and operational data.

Operational Integration: Ensuring seamless integration of predictive models with operational systems for effective resource allocation.

Future Developments:

Enhanced Modelling Techniques: Continued advancements in predictive analytics for more accurate and nuanced emission predictions.

Real-Time Adaptability: Improvements in real-time data analysis and model updating for more dynamic resource allocation.

Predictive analytics guide efficient resource allocation by identifying high-emission areas or processes, enabling targeted mitigation efforts. This approach optimizes the use of mitigation resources, leading to effective emissions reduction while maintaining operational efficiency. Strategically allocating resources based on predictive models forms a critical aspect of proactive emissions management within industries.

Benefits of Predictive Analytics in Emissions Management:

Cost Savings: Proactively managing emissions helps avoid penalties, reduce waste, and optimize resource usage, leading to cost savings.

proactive management of emissions not only benefits the environment but also offers several cost-saving advantages for industries. Here's a detailed exploration of how emissions management contributes to cost savings:

Avoidance of Penalties:

Regulatory Compliance: Proactively managing emissions ensures compliance with environmental regulations, avoiding fines or penalties for exceeding emission limits.

Avoidance of Legal Costs: Preventing legal disputes or litigations arising from non-compliance with emission regulations saves substantial legal expenses.

Waste Reduction and Efficiency:

Resource Optimization: Optimizing operations to reduce emissions often involves efficient resource utilization, leading to reduced energy and raw material waste.

Process Efficiency: Implementing emission reduction strategies often involves process improvements, enhancing overall operational efficiency and reducing waste.

Operational Optimization:

Energy Savings: Emission reduction strategies often align with energy-saving measures, leading to lower energy consumption and operational costs.

Maintenance and Upkeep: Proactive management of emissions involves regular maintenance, which prevents equipment breakdowns and reduces repair costs.

Reputation and Brand Value:

Positive Public Image: Demonstrating commitment to environmental responsibility improves brand perception, attracting environmentally conscious customers and stakeholders.

Market Advantage: A positive environmental image may open up new markets or partnerships, contributing to revenue growth and market expansion.

Supply Chain Efficiency:

Supplier Relations: Emission management initiatives that extend to the supply chain can lead to more efficient supplier relations, negotiating better terms and reducing costs.

Material Efficiency: Reducing emissions often involves optimizing material use, minimizing waste, and potentially reducing raw material costs.

Insurance and Risk Mitigation:

Insurance Premiums: Maintaining a proactive emissions management approach might lead to reduced insurance premiums due to lower associated risks.

Risk Mitigation: Minimizing environmental risks through emission management reduces the likelihood of accidents or incidents, avoiding associated costs.

Challenges and Considerations:

Initial Investment: Implementing emission reduction technologies or strategies might require initial capital investment, affecting short-term costs.

Data and Technology: Ensuring accurate emissions data and effective technology integration for emissions management might require additional investment.

Future Directions:

Cost-Effective Technologies: Continued advancements in emission reduction technologies leading to more cost-efficient solutions.

Integrated Solutions: Further integration of emission management systems with operational processes for streamlined and cost-effective management.

Proactively managing emissions not only fosters environmental stewardship but also generates tangible cost savings for industries. By avoiding penalties, reducing waste, optimizing resources, enhancing operational efficiency, and bolstering brand value, effective emission management strategies contribute significantly to overall cost reductions and improved financial performance.

Environmental Compliance: Anticipating emissions spikes facilitates compliance with environmental regulations and standards.

anticipating emissions spikes and proactively managing emissions enable industries to stay compliant with environmental regulations and standards. Here's a detailed breakdown of how anticipating emissions spikes contributes to environmental compliance:

Regulatory Adherence:

Emission Limits: Predictive analytics aid in foreseeing periods when emissions are likely to surpass regulatory thresholds, prompting proactive measures to prevent violations.

Timely Adjustments: Anticipating spikes allows industries to make timely adjustments to operations, ensuring they remain within permissible emission limits.

Preventing Non-Compliance:

Avoiding Penalties: Predicting potential emission spikes helps prevent exceeding regulatory limits, thereby avoiding fines or penalties associated with non-compliance.

Regulatory Reporting: By managing emissions proactively, industries can accurately report emissions to regulatory bodies, meeting compliance requirements.

Continuous Monitoring and Reporting:

Real-time Data Analysis: Predictive models facilitate continuous monitoring of emissions, enabling real-time assessment and adherence to regulatory standards.

Transparent Reporting: Being proactive in emission management supports transparent reporting, demonstrating commitment to compliance.

Sustainability Certifications:

Environmental Accreditations: Anticipating and managing emissions spikes align with obtaining and maintaining environmental certifications, enhancing industry reputation and compliance.

Meeting Corporate Goals: Staying ahead of emissions spikes contributes to achieving sustainability targets set by corporations, aligning with their ethical and environmental objectives.

Stakeholder Relations:

Community Engagement: Proactively managing emissions fosters positive relationships with local communities by demonstrating environmental responsibility.

Investor Confidence: Compliance with environmental regulations boosts investor confidence in the company's long-term sustainability practices.

Challenges and Considerations:

Technological Integration: Effective integration of predictive models with operational systems for timely emissions management poses a technological challenge.

Data Accuracy: The reliability of predictive models heavily depends on accurate and up-to-date emissions and operational data.

Future Developments:

Advanced Predictive Models: Continued advancements in predictive analytics for more accurate and nuanced emissions forecasting.

Regulatory Adaptation: Aligning predictive models with evolving regulatory standards to ensure ongoing compliance.

Anticipating emissions spikes through predictive analytics enables industries to proactively manage their emissions, ensuring compliance with environmental regulations and standards. This proactive approach not only avoids non-compliance penalties but also fosters positive stakeholder relations, supports sustainability certifications, and aligns with corporate environmental goals, contributing to a more responsible and compliant business operation.

Challenges and Considerations:

Data Quality: Reliable, diverse, and high-quality data is essential for accurate predictions. Incomplete or biased data can affect model accuracy.

the quality of data significantly impacts the accuracy and reliability of predictive models used for emissions forecasting. Here's an in-depth look at how data quality influences the accuracy of predictions:

Importance of Reliable Data:

Accuracy: High-quality data ensures the accuracy of predictive models, leading to more precise emissions forecasts.

Reliability: Reliable data improves the trustworthiness of predictions, aiding in effective decision-making for emissions management.

Diverse and Comprehensive Data:

Varied Data Sources: Incorporating diverse data sources (emissions, operational, weather, etc.) provides a more holistic view, enhancing prediction accuracy.

Comprehensive Parameters: Inclusion of various parameters (weather conditions, production cycles, equipment status) enriches models for more accurate predictions.

Data Completeness:

No Gaps: Complete data sets, without missing or incomplete information, ensure robustness in predictive models, reducing the risk of inaccuracies.

Temporal Consistency: Continuity and consistency in data collection over time maintain the model's effectiveness in predicting emissions.

Bias and Data Quality:

Avoiding Biases: Biased data inputs can skew predictions, affecting their accuracy. Ensuring unbiased and representative data enhances model reliability.

Data Imbalance: Addressing imbalances in data representation across different parameters ensures fair and accurate predictions.

Challenges and Mitigation:

Data Integration: Ensuring seamless integration of diverse data sources into a unified platform for accurate modelling remains a challenge.

Data Governance: Establishing robust data governance practices to maintain data quality throughout its lifecycle is essential.

Continuous Improvement:

Data Validation: Regularly validating and cleansing data to eliminate errors or inconsistencies ensures ongoing data quality.

Feedback Loop: Incorporating feedback loops to refine predictive models based on observed outcomes enhances prediction accuracy.

Future Developments:

Advanced Data Analytics: Continued advancements in data analytics to handle diverse and complex datasets for more accurate predictions.

AI-Driven Data Quality: AI-based tools to automatically identify and rectify data quality issues, ensuring more reliable predictions.

Reliable, diverse, and high-quality data forms the bedrock of accurate predictive models for emissions forecasting. Addressing issues related to completeness, diversity, biases, and consistency in data collection and integration is crucial for enhancing the accuracy and trustworthiness of predictive models, thereby enabling better-informed decisions in emissions management.

Model Validation: Continuous validation and refinement of predictive models are necessary to ensure accuracy and relevance.

Continuous validation and refinement of predictive models are crucial steps in ensuring their accuracy, reliability, and relevance in emissions forecasting. Here's an in-depth exploration of the significance of model validation and refinement:

Importance of Model Validation:

Accuracy Verification: Validation ensures that the predictive models generate accurate forecasts by comparing their outputs against real-world data.

Reliability Check: Validation assesses the reliability and consistency of model predictions under varying conditions and scenarios.

Continuous Refinement:

Feedback Integration: Incorporating feedback from observed outcomes helps refine predictive models, improving their accuracy over time.

Adaptive Learning: Models that continuously learn from new data and adjust their predictions enhance their relevance and precision.

Validation Techniques:

Cross-Validation: Using subsets of data to test the model's performance, ensuring it performs consistently across different data segments.

Out-of-Sample Testing: Evaluating the model's accuracy using data it hasn't previously encountered, verifying its predictive power.

Model Robustness:

Scenario Testing: Subjecting the model to various scenarios helps assess its robustness in predicting emissions under different conditions.

Sensitivity Analysis: Examining how changes in input variables affect model outcomes helps identify critical parameters influencing predictions.

Validation Metrics:

Accuracy Measures: Employing metrics like Mean Absolute Error (MAE), Root Mean Square Error (RMSE), or R-squared to quantify model accuracy.

Consistency Checks: Assessing how consistently the model predicts emissions over time and across different operational settings.

Challenges and Mitigation:

Data Limitations: Addressing data limitations and biases through techniques like data augmentation or sampling methods.

Model Complexity: Simplifying complex models without sacrificing accuracy for better interpretability and ease of validation.

Future Directions:

Advanced Algorithms: Integration of advanced machine learning algorithms for more precise and adaptable predictive models.

Real-Time Validation: Real-time validation techniques for immediate feedback, enhancing model accuracy promptly.

Continuous validation and refinement of predictive models are vital processes in ensuring their accuracy, reliability, and relevance in emissions forecasting. By employing various validation techniques, measuring model robustness, and incorporating feedback, industries can enhance their predictive models' accuracy, making informed decisions in emissions management and sustainability initiatives.

Future Development:

Advanced Models: Further advancements in AI and machine learning will lead to more sophisticated models capable of finer emissions predictions.

Further advancements in AI and machine learning hold great promise for the development of more sophisticated models capable of delivering finer and more accurate emissions predictions. Here's an in-depth exploration of how AI and machine learning advancements will enhance emissions prediction models:

Enhanced Predictive Capabilities:

Complex Data Patterns: Advanced AI algorithms can detect intricate patterns in diverse and extensive datasets, leading to more accurate emissions forecasts.

Non-Linear Relationships: Machine learning models excel in capturing complex non-linear relationships between various parameters affecting emissions.

Improved Model Accuracy:

Precision and Sensitivity: AI-driven models can offer higher precision and sensitivity in predicting emissions by fine-tuning parameters with more granularity.

Adaptive Learning: Machine learning models that adapt and learn from new data continuously improve their predictive capabilities over time.

Handling Big Data:

Scalability: Advanced AI models efficiently handle large and complex datasets, enabling comprehensive analysis and more accurate predictions.

Real-Time Analysis: Capabilities to process vast amounts of data in real-time allow for immediate adjustments and predictions as new data arrives.

Interdisciplinary Insights:

Integration of Diverse Data: AI can integrate data from various sources (operational, environmental, external factors) for a comprehensive emissions forecasting approach.

Multifaceted Analysis: Advanced models can consider a wider range of factors simultaneously, providing a more holistic view of emissions dynamics.

Addressing Model Complexity:

Interpretability: Advancements aim to maintain interpretability despite increased model complexity, ensuring easier understanding and validation.

Simplification Techniques: Techniques to simplify complex AI models without compromising accuracy for better model interpretation.

Future Applications:

Hybrid Models: Combining AI techniques with physics-based or statistical models for enhanced accuracy and interpretability.

Cognitive Computing: Evolution towards cognitive computing systems that mimic human thought processes for more nuanced predictions.

Ethical and Regulatory Considerations:

Ethical Use of AI: Addressing ethical concerns around AI usage, ensuring fairness, transparency, and accountability in model development and deployment.

Regulatory Compliance: Ensuring compliance with evolving regulations and standards while deploying AI-based models for emissions predictions.

Continued advancements in AI and machine learning are poised to revolutionize emissions prediction models by enabling higher accuracy, adaptability, and scalability. These sophisticated models will play a pivotal role in enabling industries to proactively manage emissions, make informed decisions, and work towards more sustainable practices while complying with regulatory standards.

Integration with Decision Support: Enhanced integration of predictive analytics with decision support systems aids in making informed decisions for emissions reduction.

The integration of predictive analytics with decision support systems is crucial in leveraging emissions predictions for informed decision-making and effective emissions reduction strategies. Here's a detailed exploration of how this integration facilitates better decision support:

Decision Support Integration:

Real-Time Insights: Integration provides real-time emissions predictions, empowering decision-makers with up-to-date information for immediate actions.

Holistic Decision-Making: Combining predictive analytics with decision support systems enables a comprehensive view, considering emissions forecasts alongside other operational parameters.

Optimized Operations:

Operational Adjustments: Decision support systems use emissions predictions to guide operational adjustments, optimizing processes for reduced environmental impact.

Resource Allocation: Insights from predictive analytics aid in resource allocation, directing efforts towards areas predicted to have higher emissions.

Risk Mitigation:

Early Warning Systems: Decision support systems equipped with emissions forecasts act as early warning systems, enabling proactive measures to mitigate potential emission spikes.

Scenario Planning: Predictive analytics within decision support systems assist in scenario planning, evaluating the impact of different decisions on future emissions.

Data-Driven Strategies:

Performance Monitoring: Integration allows decision-makers to monitor the effectiveness of implemented emission reduction strategies through real-time analytics.

Iterative Improvements: Insights from emissions predictions support iterative improvements in emission reduction tactics and strategies.

Stakeholder Engagement:

Communicating Impact: Decision support systems integrating emissions predictions facilitate transparent communication regarding the impact of decisions on emissions to stakeholders.

Strategic Planning: Incorporating emissions forecasts in strategic planning aids in aligning long-term objectives with emission reduction targets.

Challenges and Considerations:

Data Integration Complexity: Ensuring seamless integration of predictive analytics into existing decision support systems might present technical challenges.

User Training and Adoption: Adequate training and user adoption strategies are essential for maximizing the benefits of integrated systems.

Future Developments:

AI-Driven Decision Support: Advancements in AI-driven decision support systems for more intelligent, adaptive, and context-aware decision-making.

Predictive Visualization: Tools integrating predictive analytics with user-friendly visualization aids for better comprehension and decision-making.

The integration of predictive analytics with decision support systems empowers organizations to make informed, data-driven decisions in emissions reduction strategies. By providing real-time insights, optimizing operations, mitigating risks, and fostering stakeholder engagement, this integration is pivotal in aligning business decisions with environmental sustainability goals. Continual improvements in these integrated systems will further enhance their effectiveness in guiding actions towards reducing emissions and promoting sustainable practices.

AI-powered predictive analytics enable industries to forecast emissions based on historical data, weather patterns, and production schedules. This proactive approach empowers businesses to take timely actions, optimize operations, and allocate resources efficiently, contributing to better emissions management and environmental sustainability.

Optimization Algorithms: AI-driven optimization algorithms can suggest operational changes and adjustments in real-time to minimize emissions without compromising efficiency. This could involve adjusting production schedules, optimizing energy usage, or modifying processes for reduced environmental impact.

Regulatory Compliance and Reporting: Big data analytics and AI streamline the process of ensuring compliance with environmental regulations. They assist in generating accurate emissions reports, ensuring adherence to standards, and identifying areas for improvement to meet regulatory requirements.

Big data analytics and AI play a significant role in streamlining the process of ensuring regulatory compliance with environmental standards, especially in generating accurate emissions reports and identifying areas for improvement. Here's a comprehensive breakdown of their contributions:

Accurate Emissions Reporting:

Data Accuracy: Big data analytics ensure the accuracy and reliability of emissions data collected from various sources, reducing errors in reporting.

Real-time Monitoring: AI-driven systems provide real-time monitoring, enabling continuous data collection and accurate emissions reporting.

Regulatory Adherence:

Automated Compliance Checks: AI algorithms can automatically compare emissions data against regulatory thresholds, ensuring adherence to standards.

Alerts and Notifications: AI-based systems can issue alerts or notifications when emissions approach or exceed regulatory limits, prompting immediate action.

Identifying Improvement Areas:

Pattern Recognition: Big data analytics and AI identify trends and patterns in emissions data, highlighting areas requiring improvement to meet regulatory benchmarks.

Predictive Analytics: These technologies forecast future emissions trends, aiding proactive measures for compliance before potential violations occur.

Streamlined Reporting Processes:

Efficient Data Management: Big data analytics streamline the handling and processing of vast emissions datasets, facilitating accurate and timely reporting.

Automated Reporting: AI-enabled systems automate the generation of emissions reports, ensuring consistency and efficiency in regulatory submissions.

Enhancing Compliance Strategies:

Optimization Recommendations: Analytics and AI suggest optimization strategies for reducing emissions, aligning operations with regulatory standards.

Continuous Improvement: Insights derived from data analysis drive continuous improvement initiatives to meet evolving regulatory requirements.

Challenges and Considerations:

Data Privacy and Security: Safeguarding emissions data and ensuring compliance with data privacy regulations remain critical.

Interoperability: Ensuring seamless integration and interoperability of various data sources for comprehensive emissions reporting.

Future Directions:

AI-driven Compliance Solutions: Advancements in AI technologies for more adaptive and intelligent compliance monitoring and reporting.

Predictive Compliance Analytics: Further development of predictive analytics to anticipate regulatory changes and proactively adapt strategies.

Big data analytics and AI revolutionize the management of emissions data and compliance reporting by enhancing accuracy, automating processes, and identifying improvement opportunities. By leveraging these technologies, industries can ensure accurate emissions reporting, maintain regulatory adherence, and proactively implement measures for continuous improvement in meeting environmental standards and regulations.

Emissions Reduction

AI-Driven Energy Management: AI algorithms analyse energy usage patterns and recommend strategies to optimize energy consumption, thereby reducing emissions. This includes identifying opportunities for renewable energy integration and energy-efficient practices.

Optimized Supply Chain Management: AI and big data analytics optimize supply chain operations, reducing emissions associated with transportation, logistics, and production processes by identifying efficient routes, inventory management, and sustainable sourcing practices.

Continuous Improvement through Feedback Loops: AI and IoT-enabled systems create feedback loops by continuously collecting data, analysing performance, and providing actionable insights. This iterative process facilitates ongoing improvements in emissions reduction strategies.

Challenges and Future Directions:

Challenges remain, including data privacy concerns, integration of diverse data sources, and the need for standardization in data formats and collection methods. The future direction involves further advancements in AI algorithms, improved sensor technologies, and enhanced connectivity to create more robust and efficient systems for emissions monitoring, management, and optimization.

This sector is rapidly evolving, and ongoing research and development are expected to bring more innovative solutions to address environmental challenges through digital technologies.

The concept of the circular economy aims to minimize waste generation, maximize resource efficiency, and create a sustainable system by focusing on waste reduction, recycling, and sustainable resource use. Here are some trends and practices in line with circular economy principles:

Waste Reduction:

Product Design for Durability and Reusability: Manufacturers are increasingly designing products with longer lifespans and easier disassembly for repair, refurbishment, or recycling. This approach encourages consumers to choose durable and repairable products, reducing the disposal of items that are still functional.

Zero Waste Initiatives: Businesses and communities are adopting zero-waste goals, aiming to minimize landfill-bound waste through practices such as source reduction, composting, and material recovery.

Recycling:

Advanced Recycling Technologies: Innovations in recycling technologies, such as chemical recycling and advanced sorting methods using AI and robotics, are improving the efficiency and effectiveness of recycling processes. These technologies enable the recycling of a wider range of materials that were previously difficult to process.

Closed-Loop Systems: Companies are moving towards closed-loop systems where products and materials are recycled or repurposed to create new products. This approach aims to reduce reliance on virgin resources by maximizing the utilization of recycled materials.

Sustainable Resource Use:

Renewable Energy Integration: Embracing renewable energy sources reduces dependence on finite resources like fossil fuels. Businesses and governments are investing in renewable energy infrastructure to power operations and reduce their carbon footprint.

Bio-Based Materials: Utilization of bio-based materials derived from renewable sources, such as plant-based plastics, biodegradable packaging, and natural fibers, reduces reliance on petroleum-based materials and promotes sustainability.

Digital Solutions in Circular Economy:

Blockchain for Traceability: Blockchain technology is being used to create transparent supply chains, enabling consumers to trace the origins and lifecycle of products, fostering trust and accountability in recycling and sustainable practices.

AI for Waste Management: AI-powered systems are optimizing waste management processes by improving sorting efficiency in recycling facilities, predicting waste generation patterns, and identifying opportunities for waste reduction and recycling.

Policies and Collaborations:

Regulatory Support: Governments are implementing policies and regulations to incentivize circular economy practices, such as extended producer responsibility (EPR), waste reduction targets, and promoting eco-design principles.

Collaborative Initiatives: Collaboration among businesses, governments, and non-profit organizations is fostering innovation and knowledge sharing, leading to joint efforts in developing circular economy models and solutions.

These trends indicate a growing global shift towards embracing circular economy principles to create a more sustainable and resource-efficient future. Continued innovation, technological advancements, and collaborative efforts will be essential in furthering the adoption of circular economy practices across industries and communities.

Challenges and Opportunities in Achieving a Zero Carbon Future

Achieving zero carbon goals requires robust policies, international cooperation, and effective regulatory frameworks. However, several challenges hinder their implementation:

Policy Implementation Challenges:

Complexity and Long-Term Commitment: Implementing policies to achieve zero carbon involves multifaceted changes across industries, infrastructure, and society. Long-term commitment is required, but political cycles often prioritize short-term goals, leading to policy inconsistency.

Cost and Economic Impact: Transitioning to zero carbon often involves initial high costs, particularly in sectors heavily reliant on fossil fuels. Balancing economic growth with the need for sustainable policies is a challenge, as some fear potential job losses or economic disruption.

Policy Fragmentation: Inconsistent or fragmented policies across different regions or countries can hinder progress. Harmonizing policies and ensuring coherence among various levels of governance is essential for effective implementation.

INTERNATIONAL COOPERATION Challenges:

Diverse National Interests: Nations have varied energy profiles, economic priorities, and political contexts. Aligning diverse national interests towards common zero carbon goals require diplomatic negotiations and compromises.

Lack of Binding Agreements: International agreements such as the Paris Agreement set broad goals but lack binding mechanisms to ensure compliance. Some countries might not fully commit or enforce their commitments due to sovereignty concerns or differing priorities.

Technology Transfer and Support: Developing countries may lack the resources or access to clean technologies necessary for carbon reduction. Bridging this technological gap requires developed nations to support and facilitate technology transfer and capacity-building efforts.

Regulatory Framework Challenges:

Inadequate Monitoring and Enforcement: Weak monitoring mechanisms and lack of stringent enforcement can result in non-compliance and loopholes in regulations. Consistent monitoring and enforcement are crucial to ensure adherence to carbon reduction policies.

Resistance from Fossil Fuel Industries: Powerful fossil fuel industries may resist policy changes that threaten their business models. Lobbying efforts and economic interests can hinder the adoption of stringent regulations needed for carbon neutrality.

Need for Adaptability and Flexibility: Regulatory frameworks need to be adaptable to evolving technologies and changing circumstances. Rigid regulations might hinder innovation or fail to address emerging challenges effectively.

Overcoming Challenges:

Policy Alignment and Coherence: Harmonizing policies across different sectors and levels of governance is crucial.

Governments need to work collaboratively, ensuring coherence and consistency in approaches to achieve zero carbon goals.

Incentivizing Transition: Offering incentives, subsidies, and support for industries and communities transitioning to sustainable practices can mitigate economic concerns and facilitate a smoother transition.

International Dialogue and Commitment: Enhanced international dialogue, cooperation, and commitment through agreements that include binding mechanisms for accountability are vital. Greater financial and technological support for developing nations is essential for global progress.

Addressing these challenges requires sustained efforts, political will, innovative solutions, and strong leadership at national and international levels. Collaboration among governments, industries, civil society, and academia is fundamental for successful policy implementation and achieving zero carbon goals.

Transitioning to a zero-carbon economy faces several technological hurdles and research and development (R&D) challenges. Overcoming these obstacles is crucial for achieving a sustainable and decarbonized future:

Energy Storage:

Scalable and Cost-Effective Storage: Developing energy storage solutions that are both scalable to handle large capacities and cost-effective to implement at scale remains a challenge. Technologies like batteries (including lithium-ion and beyond) need further R&D to improve energy density and reduce costs.

Long-Duration Storage: Energy storage capable of providing power for extended periods, such as days or weeks, is essential for ensuring reliability in renewable energy systems. Current solutions like pumped hydro storage have limitations, and novel technologies for long-duration storage need development.

GRID INTEGRATION AND Flexibility:

Grid Resilience and Stability: Integrating intermittent renewable energy sources into existing grids poses challenges in maintaining grid stability and reliability. Developing smart grid technologies, grid-scale energy management systems, and energy demand response mechanisms are vital.

Intermittency Management: Dealing with the intermittent nature of renewable sources like solar and wind requires advanced forecasting, energy trading mechanisms, and grid-scale storage to manage fluctuations in supply and demand.

Decarbonizing Hard-to-Abate Sectors:

Heavy Industry Decarbonization: Industries like steel, cement, and chemicals, which heavily rely on high-temperature processes and fossil fuels, pose challenges in finding low-carbon alternatives. R&D is essential for developing carbon-neutral processes or alternative materials.

Aviation and Shipping: Decarbonizing transportation sectors such as aviation and shipping, which rely significantly on fossil fuels, faces hurdles in developing viable low-carbon propulsion systems, such as hydrogen fuel cells or sustainable biofuels.

CARBON CAPTURE AND Utilization (CCU):

Cost-Effective Carbon Capture Technologies: Developing cost-effective and scalable carbon capture technologies for industrial emissions and direct air capture remains a challenge. Reducing the energy requirements and cost associated with capturing and storing carbon dioxide is crucial.

Utilization of Captured CO2: Finding economically viable and environmentally sustainable uses for captured carbon dioxide, such as converting it into valuable products or storing it securely underground, requires further research and investment.

Hydrogen Economy:

Green Hydrogen Production: Generating green hydrogen through electrolysis using renewable energy sources needs R&D to improve efficiency, reduce costs, and scale up production.

Infrastructure and Storage: Establishing hydrogen infrastructure, including transportation, storage, and distribution networks, poses technological and logistical challenges.

Research and Collaboration:

Investment in R&D: Significant investment in R&D is required across multiple sectors to drive innovation and technological advancements necessary for a zero-carbon economy.

Collaborative Efforts: Collaboration among governments, industries, research institutions, and academia is vital to accelerate technological breakthroughs and overcome barriers.

Overcoming these technological hurdles and R&D challenges demands sustained investment, innovative solutions, and collaboration between various stakeholders to drive the transition to a zero-carbon economy.

Transitioning industries and infrastructure towards sustainability involves various economic challenges, costs, and significant investment needs:

Economic Challenges:

Initial Capital Investment: The shift to sustainable practices often requires substantial initial capital investments. Upgrading infrastructure, adopting new technologies, and retooling production processes can be costly for industries.

Cost Competitiveness: Sustainable alternatives might initially be more expensive than conventional practices. This cost disparity can impact the competitiveness of businesses, particularly in sectors where profit margins are thin.

Transitioning Workforce: Employment shifts may occur as some jobs in traditional industries are replaced by those in the renewable energy or sustainability sectors. Retraining and reskilling the workforce to adapt to new roles can be costly and time-consuming.

Costs and Investment Needs:

Infrastructure Upgrades: Transitioning to sustainable infrastructure, such as renewable energy generation facilities, smart grids, and electric vehicle charging networks, requires significant investment in upgrading and building new infrastructure.

Technology Development and Deployment: Research and development in green technologies, such as advanced renewable energy systems, energy-efficient machinery, and carbon capture technologies, demand substantial investment for their development and widespread deployment.

Policy Incentives and Support: Governments often need to provide financial incentives, subsidies, tax breaks, or regulatory support to encourage industries to adopt sustainable practices. These incentives may also include funding for research, development, and innovation in sustainable technologies.

Supply Chain Adaptation: Industries reliant on global supply chains may need to reconfigure their operations to source sustainable materials and components, which can involve additional costs and restructuring.

INVESTMENT OPPORTUNITIES:

Green Finance and Investment: There are growing opportunities for investments in renewable energy projects, sustainable infrastructure, clean technologies, and green bonds. Institutional investors and financial institutions are increasingly focusing on sustainability-driven investments.

Job Creation and Economic Growth: The transition to a sustainable economy can create new employment opportunities in sectors such as renewable energy, energy efficiency, green construction, and sustainable agriculture, contributing to economic growth.

Long-Term Cost Savings: While the initial investment might be high, transitioning to sustainable practices can lead to long-term cost savings. Energy-efficient technologies and renewable energy sources can reduce operational costs and mitigate risks associated with resource scarcity and climate change impacts.

Overcoming Economic Challenges:

Policy Support: Clear and consistent policies that incentivize sustainability and carbon reduction are essential. Governments can play a crucial role in providing regulatory frameworks, incentives, and subsidies to facilitate the transition.

Public-Private Partnerships: Collaboration between governments, businesses, financial institutions, and civil society is crucial to share risks, pool resources, and drive innovation for sustainable development.

Innovation and Research Funding: Increased investment in R&D for green technologies, alongside support for technology transfer and scaling, can drive down costs and make sustainable solutions more economically viable.

While transitioning industries and infrastructure towards sustainability presents economic challenges and costs, there are also significant opportunities for economic growth, job creation, and innovation. Strategic planning, targeted investments, and supportive policies can help navigate these challenges and facilitate the shift towards a more sustainable future.

Addressing social disparities, ensuring inclusivity, and engaging communities in the transition to a more sustainable future presents opportunities for fostering social equity and justice:

Opportunities for Social Equity:

Job Creation and Training: The transition to a sustainable economy offers opportunities to create new jobs in renewable energy, energy efficiency, green infrastructure, and other sectors. Training and upskilling programs can ensure that marginalized communities have access to these employment opportunities.

Energy Access and Affordability: Sustainable energy solutions can improve access to affordable and clean energy for underserved communities, reducing energy poverty and its associated health and economic burdens.

Community Ownership and Benefits: Encouraging community ownership of renewable energy projects or sustainable initiatives can ensure that the benefits and profits stay within the community, contributing to local economic development and empowerment.

Inclusivity and Diversity:

Representation and Participation: Engaging diverse voices, including those from marginalized communities, in decision-making processes and policy development ensures that their perspectives and needs are considered in the transition to a sustainable economy.

Support for Vulnerable Communities: Prioritizing support and resources for vulnerable communities, such as low-income households, indigenous groups, and communities disproportionately affected by environmental degradation or climate change, is crucial for a just transition.

Cultural Preservation and Respect: Respecting cultural heritage and traditional knowledge of indigenous communities in sustainable practices can foster inclusivity and ensure that their unique perspectives contribute to sustainable solutions.

Community Engagement:

Education and Awareness: Empowering communities through education and awareness campaigns about sustainability, climate change, and the benefits of transitioning to clean energy fosters informed decision-making and active participation.

Collaborative Partnerships: Engaging in partnerships between governments, businesses, NGOs, and local communities ensures that initiatives are community-led and address specific local needs, promoting a sense of ownership and commitment.

Equitable Access to Resources: Ensuring equitable access to resources such as funding, technology, and infrastructure enables communities to actively participate in and benefit from sustainable initiatives.

Overcoming Challenges:

Listen and Learn: Active listening and understanding the needs, concerns, and aspirations of diverse communities are essential to design inclusive and equitable policies and programs.

Empowerment and Capacity Building: Empowering communities through capacity building, training, and providing resources strengthens their ability to actively engage in and benefit from sustainable development initiatives.

Policy and Decision-Making Inclusivity: Ensuring that policies are designed with the involvement of diverse stakeholders, including community representatives, promotes inclusivity and enhances the chances of successful implementation.

A just transition to a sustainable future requires acknowledging and addressing social disparities, ensuring equitable access to opportunities, and empowering communities to actively participate in shaping their future. By leveraging community engagement and fostering inclusivity, the transition can be more effective, equitable, and sustainable for all.

Future Perspectives

Areas of Focus for Future Developments:

Predicting the precise future trajectory of technological advancements in emissions reduction and climate mitigation is challenging, but several trends and potential areas of development may shape the next generation of solutions:

Advanced Energy Storage:

Beyond Lithium-ion Batteries: Research continues to improve battery technologies, exploring alternatives such as solid-state batteries, metal-air batteries, and other novel designs that offer higher energy density, faster charging, and improved safety.

Long-Duration Energy Storage: Innovations focusing on storing renewable energy for extended periods using technologies like flow batteries, thermal storage, or hydrogen-based storage solutions are expected to gain momentum.

Carbon Capture and Utilization (CCU):

Enhanced Carbon Capture Technologies: R&D efforts aim to improve the efficiency and cost-effectiveness of carbon capture methods, including direct air capture, industrial point-source capture, and utilization or storage of captured CO_2.

Carbon Utilization Innovations: Advancements in utilizing captured CO_2 to produce valuable products such as synthetic fuels, building materials, chemicals, and even food products could play a significant role in reducing emissions.

Hydrogen Economy:

Green Hydrogen Production: Further advancements in electrolysis methods using renewable energy sources to produce green hydrogen are anticipated, leading to increased efficiency, reduced costs, and scaled-up production.

Hydrogen Applications: The expansion of hydrogen fuel cell technologies for various sectors, including transportation, industry, and energy storage, is expected to grow, contributing to emissions reduction.

Artificial Intelligence and Big Data:

AI-Driven Solutions: More sophisticated AI algorithms and machine learning models will enhance emissions forecasting, optimize energy systems, and improve the efficiency of renewable energy integration into grids.

Big Data Analytics: Advanced data analytics will play a crucial role in identifying emission hotspots, predicting climate-related risks, and optimizing resource utilization in various industries.

Innovative Materials and Processes:

Green Chemistry: Research into sustainable materials and chemical processes that reduce emissions or use renewable resources as alternatives to conventional materials will continue to expand.

Cleantech Innovation: Innovations in various sectors, including agriculture, transportation, construction, and waste management, will focus on developing cleaner technologies and practices to reduce environmental impact.

Collaboration and Policy Support:

International Cooperation: Increased collaboration among nations, industries, and research institutions will accelerate the development and deployment of innovative climate solutions.

Policy Frameworks: Supportive policies, regulations, and incentives from governments worldwide will be crucial in driving investments, fostering innovation, and incentivizing the adoption of low-carbon technologies.

The future trajectory of technological advancements in emissions reduction and climate mitigation will likely involve a convergence of multiple innovative solutions across various sectors. Continued R&D investments, collaboration among stakeholders, supportive policies, and a focus on sustainable innovation will be key drivers in shaping the next generation of climate solutions.

Potential shifts in global climate agreements, policies, and multilateral cooperation toward more ambitious emissions reduction targets are essential for addressing the pressing challenges of climate change. Several trends and developments may influence these shifts:

Enhanced Climate Ambition:

Strengthened National Commitments: Countries may enhance their Nationally Determined Contributions (NDCs) under the Paris Agreement, setting more ambitious targets for emissions reduction, renewable energy adoption, and adaptation measures.

Net-Zero Commitments: More nations, regions, cities, and businesses are expected to announce net-zero emissions targets, aiming to achieve carbon neutrality by mid-century or earlier, signalling a collective push for ambitious climate action.

Increased International Cooperation:

COP Negotiations and Agreements: The Conference of the Parties (COP) meetings, including the UN Climate Change Conference (COP26, COP27, etc.), are crucial platforms for countries to negotiate and update their climate commitments, potentially leading to more ambitious agreements.

Collaborative Initiatives: Enhanced cooperation among nations, such as joint commitments, technology sharing, financial aid to developing countries, and climate-related partnerships, can drive collective efforts towards emissions reduction and adaptation.

Financial Commitments and Support:

Climate Finance: Increased commitments from developed countries to provide financial support to developing nations for climate adaptation, mitigation, and capacity-building efforts will be crucial for global climate efforts.

Private Sector Engagement: Encouraging private investment in sustainable projects, green technologies, and climate-resilient infrastructure through incentives, regulations, and public-private partnerships will complement public funding.

Regulatory and Policy Frameworks:

Carbon Pricing Mechanisms: Expansion and strengthening of carbon pricing mechanisms, such as carbon taxes or cap-and-trade systems, may gain traction to incentivize emissions reduction across various sectors.

Energy Transition Policies: Governments implementing ambitious policies to accelerate the transition from fossil fuels to renewable energy sources, including phasing out subsidies for fossil fuels and promoting clean energy investments.

Adaptation and Resilience Building:

Focus on Climate Resilience: Increasing emphasis on adaptation measures, such as investing in climate-resilient infrastructure, agriculture, and disaster risk reduction, to mitigate the impacts of climate change.

Nature-Based Solutions: Greater recognition and investment in nature-based solutions, such as reforestation, ecosystem restoration, and sustainable land use practices, for carbon sequestration and adaptation.

Youth and Civil Society Engagement:

Public Mobilization: Growing demands from youth-led movements, civil society, and grassroots initiatives for governments and policymakers to take stronger action on climate change could drive more ambitious policies.

Accountability and Transparency: Increased scrutiny and calls for accountability of governments and corporations to align their actions with climate commitments, fostering a culture of transparency in climate actions.

These potential shifts toward more ambitious emissions reduction targets and enhanced multilateral cooperation require sustained political will, collaborative efforts, and comprehensive policy frameworks at both national and international levels. Continued engagement, innovation, and commitment from all stakeholders will be crucial in achieving a more sustainable and resilient future.

Education, public awareness, and behavioural changes play a pivotal role in shaping a sustainable future. Forecasting their role involves recognizing trends that might influence how these factors contribute to sustainability:

Education and Awareness:

Integrated Sustainability Education: There's a growing trend toward integrating sustainability education into school curricula at all levels, fostering awareness, knowledge, and values related to environmental stewardship and social responsibility.

Lifelong Learning Initiatives: Continued education and awareness campaigns aimed at adults, workplaces, and communities can foster ongoing learning about sustainability issues, encouraging informed decision-making and behavioural changes.

Behavioural Changes:

Shift Towards Sustainable Lifestyles: Increasing awareness of environmental and social issues might lead to a shift in consumer behaviour, promoting sustainable consumption patterns, minimalism, and conscious purchasing decisions.

Adoption of Greener Practices: Awareness campaigns and education can encourage individuals and communities to adopt greener habits, such as reducing waste, conserving energy and water, and using alternative modes of transportation.

Technology and Communication:

Digital Platforms for Education: Advancements in technology enable broader access to sustainability-related information and educational resources through online platforms, mobile apps, and social media, fostering greater awareness and engagement.

Social Media and Influencer Engagement: Influencers and social media platforms can play a significant role in promoting sustainable behaviours, leveraging their reach to advocate for environmental consciousness and lifestyle changes.

Community Engagement:

Local Action and Grassroots Initiatives: Increasing participation in community-driven sustainability initiatives, local activism, and collaborative projects aimed at addressing environmental and social challenges at a grassroots level.

Participatory Decision-Making: Involving communities in decision-making processes related to sustainability policies, urban planning, and environmental projects fosters a sense of ownership and commitment to collective action.

Future Projections:

Empowerment Through Information: Continued efforts to disseminate accurate and accessible information about climate change, biodiversity loss, and social justice issues will empower individuals to take informed action.

Cultural Shifts and Norms: Over time, sustained education and awareness campaigns may lead to cultural shifts, where sustainable practices become societal norms and values, influencing broader behavioural changes.

Challenges and Opportunities:

Addressing resistance to change, especially regarding adopting sustainable practices, involves strategies aimed at education, storytelling, and illustrating the benefits of sustainable living. Here's how these approaches can effectively overcome scepticism:

Education and Awareness:

Clear Communication of Facts: Providing accurate and easy-to-understand information about the impacts of climate change, environmental degradation, and the benefits of sustainable living fosters awareness and understanding.

Highlighting Long-Term Benefits: Emphasizing the long-term benefits of sustainable practices, such as cost savings, improved health, enhanced quality of life, and reduced environmental impact, can motivate individuals to embrace change.

Storytelling and Engagement:

Narratives and Personal Stories: Sharing relatable stories and experiences of individuals or communities who have successfully adopted sustainable practices can inspire and connect with audiences on a personal level.

Showcasing Success Stories: Highlighting successful examples of businesses, cities, or countries that have transitioned to sustainable models effectively demonstrates that change is feasible and beneficial.

Making Sustainability Tangible:

Practical Examples and Demonstrations: Providing tangible demonstrations or hands-on experiences showcasing the ease and effectiveness of sustainable technologies or practices can dispel doubts and encourage adoption.

Visualizations and Simulations: Utilizing visual aids, simulations, or virtual experiences to illustrate the immediate and long-term impacts of unsustainable behaviours versus sustainable alternatives can be persuasive.

Engaging Stakeholders:

Involving Communities in Solutions: Encouraging community involvement in finding solutions, allowing them to participate in decision-making processes, and addressing their concerns builds trust and fosters ownership of change.

Collaborative Partnerships: Collaborating with diverse stakeholders, including governments, businesses, NGOs, and local communities, can amplify messages and initiatives, creating a unified front for change.

Empathy and Inclusivity:

Understanding Concerns and Perspectives: Listening to and addressing the concerns and perspectives of individuals resistant to change fosters empathy, understanding, and tailoring approaches to better meet their needs.

Celebrating Progress: Recognizing and celebrating small victories and progress towards sustainability encourages positive reinforcement and motivates continued action.

Promoting Behavioural Nudges:

Making Sustainable Choices Easier: Implementing subtle nudges through design, incentives, or default choices that make sustainable options more accessible, convenient, and attractive.

Social Norms and Peer Influence: Leveraging social influence by showcasing those embracing sustainable behaviours is becoming a societal norm and encouraging positive peer pressure.

Addressing resistance to change requires multifaceted approaches that combine education, storytelling, relatable examples, and engaging experiences to shift mindsets and behaviours. Tailoring strategies to address specific concerns and motivations of individuals or communities is key to fostering a broader acceptance and adoption of sustainable practices.

Harnessing emerging technologies offers vast opportunities to create innovative educational tools, interactive learning experiences, and engagement platforms that can inspire action and cultivate a sense of global citizenship toward sustainability:

Virtual Reality (VR) and Augmented Reality (AR):

Immersive Learning Environments: VR and AR can create immersive experiences that transport users to various environments, allowing them to witness the impacts of climate change firsthand or experience sustainable practices in different settings.

Interactive Simulations: Simulations through VR or AR can demonstrate cause-and-effect scenarios related to environmental changes, enabling users to explore solutions and understand complex issues in a more engaging manner.

Gamification and Serious Games:

Educational Gaming: Gamified platforms can provide entertaining yet educational experiences where users solve sustainability-related challenges, make decisions, and witness the consequences of their actions, fostering learning through play.

Collaborative Problem-Solving: Games that encourage collaboration among players to address real-world sustainability issues, such as managing resources or building sustainable communities, promote teamwork and problem-solving skills.

Mobile Apps and Interactive Tools:

Sustainable Lifestyle Apps: Mobile applications that track personal carbon footprints, suggest eco-friendly alternatives, and provide tips for sustainable living empower users to make informed choices in their daily lives.

Interactive Tools for Action: Apps or platforms that facilitate community engagement, organize sustainability-focused events, and connect individuals with local initiatives can encourage collective action and engagement.

Online Platforms and social media:

Digital Engagement Platforms: Online forums, social media groups, and interactive websites dedicated to sustainability discussions, idea sharing, and showcasing success stories encourage active participation and knowledge exchange.

Crowdsourced Initiatives: Platforms that enable crowdsourcing of ideas, solutions, or projects related to sustainability can foster a sense of belonging and empower individuals to contribute to larger global goals.

Artificial Intelligence (AI) and Personalization:

Personalized Learning Experiences: AI-powered educational tools can adapt content based on individual learning styles, preferences, and progress, creating tailored experiences for more effective learning.

Predictive Analytics for Behaviour Change: AI algorithms analysing user behaviour can provide personalized recommendations, nudges, or feedback to encourage sustainable actions and habits.

Remote Learning and Accessibility:

Remote Education Platforms: Leveraging online education and remote learning technologies ensures accessibility and scalability of sustainability education to a global audience, irrespective of geographical limitations.

Inclusive and Multilingual Content: Creating content in multiple languages and ensuring inclusivity for diverse cultures and communities enhances the reach and impact of educational tools.

Leveraging these emerging technologies in innovative ways can transform education, engagement, and awareness-building efforts for sustainability. By creating engaging, interactive, and personalized experiences, these tools can inspire action, empower individuals, and cultivate a sense of global responsibility towards creating a more sustainable future.

forecasting the role of education, public awareness, and behavioural changes underscores their immense potential in shaping a sustainable future:

Education as a Catalyst:

Empowering Future Generations: Education instils values, knowledge, and critical thinking skills necessary for individuals to understand complex environmental challenges and actively participate in shaping a sustainable future.

Lifelong Learning for Adaptation: Continuous education, throughout life, equips people with updated information, new skills, and adaptive strategies needed to navigate and respond to evolving sustainability issues.

Public Awareness for Engagement:

Building Consciousness and Responsibility: Increased public awareness fosters a sense of responsibility and ownership toward environmental issues, motivating individuals to make informed choices and advocate for change.

Creating a Culture of Sustainability: A well-informed public can influence societal norms, encouraging sustainable practices and influencing businesses and policymakers to prioritize environmental considerations.

Behavioural Changes Driving Action:

Individual and Collective Impact: Behavioural changes, such as adopting sustainable lifestyles, reducing waste, conserving resources, and advocating for policy changes, collectively contribute to mitigating environmental degradation.

Influencing Policy and Industry: Societal shifts in behaviours and demands influence policy decisions and industries, compelling them to align with more sustainable practices and adapt to changing consumer preferences.

Societal Attitudes and Policies:

Cultural Shifts for Sustainability: Education and awareness initiatives can catalyse cultural shifts where sustainability becomes a societal value, influencing social norms and encouraging responsible behaviours.

Policy Support and Collaboration: Heightened public awareness and behavioural changes create a conducive environment for policymakers to implement more ambitious policies, incentivize sustainability, and collaborate across sectors.

Continued Efforts for Impact:

Sustaining Momentum: Continuous efforts in education, awareness, and behavioural change campaigns sustain the momentum for sustainable practices, ensuring long-term commitment and action.

Adapting to Challenges: As new challenges arise, ongoing education and awareness efforts adapt, providing the necessary information and tools to address emerging sustainability issues effectively.

The collective impact of education, public awareness, and behavioural changes can drive a profound transformation towards a more environmentally conscious and equitable world. These factors synergistically reinforce each other, laying the groundwork for societal attitudes, behaviours, and

181

policies that prioritize sustainability and contribute to a better future for generations to come.

Understanding the Basics of Carbon Emissions

―――

Carbon emissions refer to the release of carbon compounds, primarily carbon dioxide (CO_2), into the atmosphere due to human activities like burning fossil fuels, deforestation, industrial processes, and agriculture. These emissions trap heat in the Earth's atmosphere, contributing to global warming and climate change.

Historical Evolution of Carbon Emissions Research

Pre-Industrial Revolution: Minimal human-induced carbon emissions.

Industrial Revolution: Accelerated use of coal, oil, and gas led to a substantial increase in carbon emissions.

20th Century: Advancements in technology further increased emissions.

Late 20th Century to Present: Growing awareness of climate change; significant research into carbon emissions' impact and mitigation strategies.

Significance of Carbon Emissions in Today's World

Carbon emissions are a pressing global issue due to their role in climate change, which poses threats like rising temperatures, extreme weather events, sea-level rise, biodiversity loss, and disruptions to ecosystems and economies. Addressing carbon emissions is crucial to mitigating these risks.

Current Global Carbon Emission Trends

Rising Emissions: Despite efforts to reduce emissions, global carbon emissions have continued to rise due to increasing industrialization, urbanization, and energy demand.

Sectors Contributing: Energy production, transportation, industry, agriculture, and deforestation are major contributors to carbon emissions.

International Agreements and Initiatives: Various international agreements (e.g., Paris Agreement) aim to limit global warming by reducing carbon emissions. Countries set targets and implement policies to curb emissions.

Sources of Carbon Emissions

1. Fossil Fuels and Carbon Emissions

Coal, Oil, and Natural Gas: Burning fossil fuels for energy production in power plants, transportation, heating, and manufacturing releases significant amounts of carbon dioxide (CO_2) into the atmosphere.

Electricity Generation: Power plants burning coal and natural gas are among the largest sources of CO_2 emissions globally.

———

2. INDUSTRIAL PROCESSES and Emissions

Manufacturing: Industries like cement, steel, chemicals, and refining contribute to carbon emissions due to the energy-intensive processes involved.

Chemical Reactions: Some industrial processes produce emissions as a byproduct of chemical reactions, leading to the release of greenhouse gases like methane (CH_4) and nitrous oxide (N_2O).

3. Agriculture and Land Use Related Emissions

Livestock Farming: Enteric fermentation in ruminant animals (such as cows and sheep) generates methane. Manure management also produces methane and nitrous oxide.

Deforestation: Clearing forests for agriculture and other purposes releases stored carbon and reduces the planet's capacity to absorb CO_2.

4. Transportation and its Carbon Footprint

Vehicular Emissions: Cars, trucks, ships, airplanes, and other forms of transportation burning fossil fuels release CO_2 and other pollutants.

Infrastructure and Operations: The overall carbon footprint of transportation also includes emissions from building and maintaining transportation infrastructure, such as roads, airports, and ports.

Impact and Mitigation Efforts

Impact: These sources collectively contribute to the increasing concentration of greenhouse gases in the atmosphere, leading to global warming and climate change.

Mitigation Efforts: Strategies include transitioning to renewable energy, improving energy efficiency, implementing sustainable agricultural practices, promoting public transportation, and investing in low-carbon technologies.

Understanding and addressing these sources of carbon emissions are vital in global efforts to mitigate climate change and transition to a more sustainable future.

Impact of Carbon Emissions

1. Climate Change and Global Warming

Rising Temperatures: Increased concentration of greenhouse gases, including carbon dioxide, leads to a warming effect on the planet, altering weather patterns and causing heatwaves, droughts, and more extreme weather events.

Glacial Melting and Sea-Level Rise: Higher temperatures contribute to the melting of glaciers and ice caps, leading to rising sea levels, which pose a threat to coastal communities and ecosystems.

Shifts in Ecosystems: Climate change alters ecosystems, affecting plant and animal species' distribution, migration patterns, and habitat availability.

2. Effects of Carbon Emissions on Biodiversity

Loss of Habitat: Deforestation and changing climate conditions due to carbon emissions threaten biodiversity by reducing habitats for many species.

Species Extinction Risk: Some species may face an increased risk of extinction due to climate-related changes in their habitats and ecosystems.

3. Ocean Acidification and Carbon Emissions

Carbon Absorption: Oceans absorb a significant portion of atmospheric carbon dioxide, leading to increased acidity in seawater.

Impact on Marine Life: Acidification affects marine ecosystems, particularly calcifying organisms like corals, mollusks, and some plankton species, threatening their ability to build shells or skeletons.

4. Societal Impact and Public Health

Extreme Weather Events: Increased frequency and severity of extreme weather events due to climate change can result in property damage, displacement, and loss of life.

Health Risks: Air pollution from carbon emissions contributes to respiratory problems, cardiovascular diseases, and other health issues, particularly in urban areas with high pollution levels.

Mitigation and Adaptation Efforts

Mitigation Strategies: Transitioning to renewable energy, improving energy efficiency, reforestation, and adopting sustainable practices across sectors.

Adaptation Measures: Building resilient infrastructure, developing early warning systems for extreme weather events, and implementing policies to address health risks associated with climate change.

Understanding the multifaceted impacts of carbon emissions is crucial in formulating effective policies and strategies to mitigate climate change and minimize its adverse effects on the environment, biodiversity, and human well-being.

Measuring Carbon Emissions

Measuring carbon emissions involves various methodologies, standards, tools, and technologies to quantify the amount of greenhouse gases released into the atmosphere. Here's an overview:

1. Carbon Foot printing Methods and Standards

Carbon Footprint: It's a measure of the total greenhouse gas emissions caused directly or indirectly by an individual, organization, event, or product.

Standards: Several standards exist for calculating carbon footprints, such as the GHG Protocol, ISO 14064, PAS 2050, and others, providing guidelines and frameworks for measurement and reporting.

2. Tools and Technologies for Measuring Emissions

Emission Inventories: Governments and organizations use emission inventories to estimate emissions from various sectors (e.g., energy, industry, transportation, agriculture) using data on fuel consumption, production processes, and more.

Remote Sensing: Satellite-based monitoring and remote sensing technologies help in measuring emissions from large geographical areas, such as monitoring deforestation or tracking changes in carbon sinks.

Emission Monitoring Systems: Industrial facilities use real-time monitoring systems to track and report their emissions, employing sensors and data collection tools to measure and manage their carbon output.

3. Data Collection and Analysis Techniques

Direct Measurement: Using instruments and sensors to directly measure emissions from sources like power plants, vehicles, and industrial facilities.

Estimation and Modelling: Employing models and algorithms based on available data to estimate emissions when direct measurement is not feasible.

Life Cycle Assessment (LCA): Analysing the entire life cycle of a product or process to assess its environmental impact, including carbon emissions from raw material extraction to disposal.

Challenges and Advances

Data Accuracy and Availability: Access to accurate and comprehensive data remains a challenge for precise carbon footprint assessments.

Advances in Technology: Continuous advancements in sensor technology, data analytics, and remote sensing enhance the accuracy and efficiency of measuring emissions.

Integration of Standards: Efforts to harmonize and standardize measurement methodologies facilitate consistency and comparability of carbon footprint assessments globally.

Importance

Decision-Making: Accurate measurement of carbon emissions is crucial for policymakers, businesses, and individuals to make informed decisions regarding emission reduction strategies and resource allocation.

Accountability and Reporting: Measuring emissions allows entities to be accountable for their environmental impact and report their progress in reducing carbon footprints.

Measuring carbon emissions using standardized methodologies and advanced technologies is essential for effective climate change mitigation, fostering a better understanding of our impact on the environment, and implementing strategies to reduce greenhouse gas emissions.

Here are some key mitigation strategies aimed at reducing carbon emissions and combating climate change:

Mitigation Strategies

1. Renewable Energy Sources

Solar Power: Harnessing energy from the sun through photovoltaic panels or concentrated solar power systems.

Wind Energy: Using wind turbines to generate electricity from wind power.

Hydropower: Generating electricity from flowing water in rivers or dams.

Biomass and Bioenergy: Utilizing organic materials, such as agricultural residues or organic waste, to produce energy.

2. Energy Efficiency and Conservation Techniques

Improved Technologies: Using energy-efficient appliances, LED lighting, and smart energy management systems to reduce energy consumption.

Building Design: Constructing energy-efficient buildings with better insulation, efficient HVAC systems, and sustainable materials.

Transportation: Promoting fuel-efficient vehicles, public transportation, and alternative fuels to decrease emissions from transportation.

3. Carbon Capture and Storage (CCS)

Capture: Technology to capture carbon dioxide emissions from industrial processes or power plants before it reaches the atmosphere.

Transportation: Transporting the captured CO_2 to storage sites.

Storage: Storing CO_2 in geological formations like depleted oil and gas fields or deep saline aquifers, preventing its release into the atmosphere.

4. Afforestation and Reforestation Initiatives

Afforestation: Planting trees in areas that were previously devoid of forest cover.

Reforestation: Restoring and replanting trees in areas that have been deforested or degraded.

Forest Management: Sustainable forest management practices to conserve existing forests and enhance their carbon sequestration capacity.

Importance and Challenges

Reducing Emissions: These strategies aim to significantly reduce carbon emissions by shifting to cleaner energy sources, minimizing energy waste, and actively capturing and storing carbon.

Challenges: Cost-effectiveness, technological advancements, policy implementation, and public acceptance are among the challenges in deploying these strategies on a large scale.

Synergistic Approach

Combining multiple mitigation strategies often yields more significant benefits. For instance, integrating renewable energy with energy-efficient practices or using afforestation alongside carbon capture technologies can create a more comprehensive approach to reducing carbon emissions.

Implementing these mitigation strategies globally is crucial to limit the rise in global temperatures and mitigate the impacts of climate change.

1. National and International Climate Policies

National Policies:

Renewable Energy Targets: Many countries set targets for increasing the share of renewable energy in their energy mix.

Energy Efficiency Standards: Implementing regulations and standards to improve energy efficiency in industries, transportation, and buildings.

Emissions Reduction Targets: Setting specific goals to reduce greenhouse gas emissions within a specified timeframe.

International Agreements:

The Paris Agreement: A landmark international treaty adopted in 2015 under the United Nations Framework Convention on Climate Change (UNFCCC). It aims to limit global warming to well below 2 degrees Celsius and pursue efforts to limit it to 1.5 degrees Celsius compared to pre-industrial levels.

2. The Paris Agreement and its Implications

Key Aspects:

Nationally Determined Contributions (NDCs): Each participating country sets its own targets and action plans (NDCs) to mitigate climate change based on its capabilities and circumstances.

Global Cooperation: The agreement emphasizes global collaboration and support for developing countries in their efforts to mitigate and adapt to climate change.

Transparency and Accountability: Regular reporting and assessment mechanisms ensure transparency and accountability among nations regarding their emission reduction efforts.

Implications:

Long-Term Goals: Encourages long-term strategies to achieve climate resilience and low-carbon development pathways.

Mobilization of Resources: Aims to mobilize financial resources, technology transfer, and capacity-building efforts to assist developing countries in their climate action plans.

3. Carbon Pricing and Emission Trading Systems

Carbon Pricing:

Carbon Tax: A direct tax levied on the carbon content of fossil fuels or on emissions.

Emission Trading Systems (ETS): Cap-and-trade systems set a limit on emissions and allow entities to buy and sell emission allowances.

Objectives and Challenges:

Incentivizing Emission Reduction: Carbon pricing aims to create economic incentives for reducing emissions by putting a price on carbon.

Complex Implementation: Designing and implementing effective carbon pricing mechanisms face challenges related to setting appropriate prices, ensuring fairness, and addressing competitiveness concerns.

4. Regulatory Frameworks for Emissions Control

Government Regulations:

Emission Standards: Setting limits on allowable emissions from industries, vehicles, and power plants.

Environmental Policies: Implementing policies that encourage the use of cleaner technologies and penalize excessive emissions.

Importance:

Compliance and Enforcement: Regulatory frameworks ensure compliance with emission standards and help enforce penalties for non-compliance.

Public Health and Environmental Protection: These regulations aim to protect public health and the environment by controlling harmful emissions.

These policies, agreements, pricing mechanisms, and regulatory frameworks play critical roles in shaping global efforts to mitigate climate change and transition towards a more sustainable, low-carbon future.

Economic Costs of Carbon Emissions

Climate-Related Costs:

Direct Costs: Damages from extreme weather events, sea-level rise, and agricultural losses due to changing climate patterns.

Indirect Costs: Impact on infrastructure, health care, insurance, and the overall economy due to climate-related disruptions.

Mitigation Costs:

Transition Costs: Investments required to transition from fossil fuels to renewable energy and low-carbon technologies.

Adaptation Costs: Investments in building resilience against climate change impacts, such as developing infrastructure to withstand extreme weather events.

2. Social Equity and Climate Justice

Vulnerable Communities:

Disproportionate Impacts: Marginalized communities often bear the brunt of climate change impacts due to inadequate resources, infrastructure, and social support systems.

Environmental Justice: Addressing the unequal distribution of environmental benefits and burdens, ensuring fair treatment and involvement of all people in environmental policies and decision-making.

Climate Refugees and Displacement:

Displacement: Rising sea levels, extreme weather events, and environmental degradation can force people to migrate, leading to climate-induced displacement and refugees.

3. Job Creation in the Green Economy

Green Jobs:

Renewable Energy Sector: Employment opportunities in solar, wind, hydro, and other renewable energy industries.

Energy Efficiency: Jobs in retrofitting buildings, installing energy-efficient systems, and developing green technologies.

Conservation and Restoration: Employment in afforestation, reforestation, and ecosystem restoration initiatives.

Skills Transition:

Retraining Programs: Initiatives to train workers from traditional carbon-intensive industries for roles in the green economy.

Investment in Innovation: Promoting research and development in clean energy and sustainable technologies to create new job opportunities.

Importance and Challenges

Economic Opportunities:

Long-Term Economic Benefits: Transitioning to a low-carbon economy can create new markets, enhance innovation, and lead to sustainable economic growth.

Social Co-Benefits: Green initiatives often bring improved health, increased energy security, and reduced pollution.

Challenges:

Equitable Transition: Ensuring that the shift to a green economy benefits all sectors of society, especially vulnerable communities and workers in carbon-intensive industries.

Investment Needs: Significant investments and policy support are required to facilitate the transition and address social and economic challenges.

Addressing economic and social aspects related to carbon emissions and climate change requires a comprehensive approach that considers both mitigation and adaptation strategies while ensuring social equity and inclusive economic growth. Efforts towards sustainability should prioritize addressing the needs of vulnerable communities and creating opportunities for a just transition to a green economy.

INNOVATIONS AND TECHNOLOGY play a crucial role in addressing carbon emissions and achieving sustainability goals. Here's an overview of clean technologies, the role of AI and big data, and the concept of smart cities and sustainable infrastructure:

1. Clean Technologies and Innovations

Renewable Energy Advancements:

Solar Power: Innovations in solar photovoltaic technology, including improvements in efficiency and cost reduction.

Wind Energy: Advancements in wind turbine design and offshore wind farms for increased energy production.

Battery Storage: Developing better energy storage solutions to address intermittency issues in renewable energy sources.

Energy Efficiency and Smart Grids:

Smart Meters and Grids: Integrating sensors and data analytics to optimize energy distribution and reduce wastage.

Building Technologies: Innovations in smart building systems for improved energy efficiency and reduced carbon footprint.

2. Role of AI and Big Data in Emissions Reduction

AI Applications:

Emissions Monitoring: AI-powered systems for real-time monitoring and analysis of emissions from industries, transportation, and power plants.

Energy Optimization: AI algorithms to optimize energy usage in various sectors, predicting demand, and enhancing efficiency.

Climate Modeling: Utilizing AI and big data to improve climate modeling, predict climate patterns, and assess risks.

Big Data Analytics:

Data-Driven Decision Making: Analyzing vast datasets to identify emission hotspots, track progress, and inform policy decisions.

Supply Chain Optimization: Using big data analytics to optimize supply chains, reduce waste, and enhance sustainability.

3. Smart Cities and Sustainable Infrastructure

Sustainable Urban Planning:

Smart Transportation: Integrating technologies like electric vehicles, intelligent traffic management, and public transit systems to reduce emissions.

Green Buildings: Implementing sustainable building designs, utilizing renewable energy sources, and optimizing resource use.

Digital Infrastructure:

IoT and Connectivity: Creating interconnected systems using the Internet of Things (IoT) for efficient resource management and energy conservation.

Waste Management: Utilizing smart technologies for efficient waste collection, recycling, and reducing landfill emissions.

Impact and Challenges

Positive Impact:

Efficiency Gains: Implementing innovative technologies leads to reduced energy consumption, lower emissions, and improved resource management.

Quality of Life: Smart city initiatives can enhance the quality of life by improving air quality, transportation, and overall sustainability.

Challenges:

Integration and Scale: Scaling up innovative technologies across cities and regions while ensuring interoperability and seamless integration.

Equity and Accessibility: Ensuring that technological advancements benefit all communities and don't exacerbate existing inequalities.

Innovations in clean technologies, the integration of AI and big data, and the development of smart cities and sustainable infrastructure are crucial in the global effort to reduce carbon emissions, promote sustainability, and build a more resilient future.

Corporate responsibility plays a significant role in addressing carbon emissions and promoting sustainability. Here's an overview of corporate sustainability, emissions reduction efforts, supply chain management, and carbon neutrality initiatives:

1. Corporate Sustainability and Emissions Reduction

Setting Emission Reduction Goals:

Emission Targets: Many companies set specific targets to reduce their carbon footprint, often aligning with international goals (e.g., Science-Based Targets Initiative).

Renewable Energy Adoption: Companies invest in renewable energy sources to power their operations and reduce reliance on fossil fuels.

Energy Efficiency and Innovation:

Efficiency Programs: Implementing energy-saving initiatives within operations, such as upgrading equipment, improving processes, and optimizing energy use.

Innovative Technologies: Investing in and adopting new technologies that reduce emissions in manufacturing, transportation, and other operational aspects.

2. Supply Chain Management and Carbon Emissions

Supply Chain Sustainability:

Supplier Engagement: Encouraging suppliers to adopt sustainability practices, reduce emissions, and provide environmentally friendly products and services.

Traceability and Transparency: Implementing systems to trace the environmental impact of products throughout the supply chain.

Carbon Accounting and Offsetting:

Carbon Accounting: Measuring emissions across the supply chain to identify hotspots and areas for improvement.

Offsetting Strategies: Investing in carbon offset projects to compensate for unavoidable emissions, such as reforestation or renewable energy projects.

3. Carbon Neutrality and Corporate Initiatives

Achieving Carbon Neutrality:

Net-Zero Emissions Commitments: Companies pledge to achieve net-zero emissions by a certain date, often by reducing emissions and offsetting remaining emissions.

Carbon Offsetting: Investing in projects that reduce or remove carbon emissions equivalent to the company's emissions.

Corporate Initiatives:

Circular Economy Practices: Embracing circular economy principles to minimize waste and maximize the use of resources.

Stakeholder Engagement: Engaging with stakeholders, including investors, employees, and communities, to align sustainability efforts and ensure accountability.

Importance and Challenges

Business Imperative:

Reputation and Brand Image: Companies committed to sustainability often have a better reputation and appeal to environmentally conscious consumers.

Risk Mitigation: Addressing environmental concerns helps mitigate regulatory risks and adapt to changing market demands.

Challenges:

Complex Supply Chains: Managing emissions across global supply chains presents challenges in data collection, standardization, and cooperation with multiple partners.

Costs and Investments: Implementing sustainable practices may require significant investments and sometimes poses financial challenges in the short term.

Corporate responsibility and sustainability efforts play a crucial role in mitigating climate change and reducing carbon emissions. Companies are increasingly recognizing the importance of aligning their operations with environmental goals and making substantial commitments to achieve carbon neutrality and promote a more sustainable future.

Behavioural change, education, public awareness, and community engagement are essential components in achieving emissions reduction and fostering a more sustainable society. Here's an overview of their significance:

1. Public Awareness and Behavioural Changes

Importance of Awareness:

Informing about Climate Change: Increasing public awareness about the causes and consequences of climate change encourages people to take action.

Motivating Action: Educating individuals about their carbon footprint and the impact of their daily choices can drive behavioural changes.

Influencing Behaviours:

Energy Consumption: Encouraging energy conservation practices, such as reducing electricity usage and using energy-efficient appliances.

Transportation: Promoting public transit, carpooling, biking, or using electric vehicles to reduce emissions from transportation.

Waste Reduction: Advocating for waste reduction, recycling, and composting to minimize landfill emissions.

2. Education and Climate Literacy

Importance of Education:

Climate Literacy: Teaching individuals, from school children to adults, about climate science, environmental issues, and sustainable practices.

Empowering Action: Providing knowledge and skills to make informed decisions and advocate for sustainability in various aspects of life.

Educational Initiatives:

School Curriculum: Integrating climate change and sustainability topics into school curricula at various levels of education.

Awareness Campaigns: Conducting outreach programs, workshops, and seminars to educate communities about environmental issues.

3. Community Engagement for Emissions Reduction

Grassroots Initiatives:

Community Programs: Engaging local communities in initiatives like tree planting, clean-up drives, and local sustainability projects.

Collaborative Efforts: Partnering with community organizations, NGOs, and local governments for collective action on emissions reduction.

Behavioural Shifts:

Social Norms and Influence: Leveraging social networks and community leaders to promote sustainable behaviours and influence social norms.

Inclusive Engagement: Encouraging participation from diverse groups within communities to ensure inclusive decision-making and actions.

Impact and Challenges

Impact of Behavioural Changes:

Cumulative Effect: Small individual changes in behavior, when adopted widely, can collectively lead to significant emissions reductions.

Long-Term Sustainability: Encouraging sustainable behaviours fosters long-term habits that contribute to a more sustainable future.

Challenges:

Overcoming Inertia: Changing established behaviours can be challenging and may require incentives, education, and sustained efforts.

Equity and Access: Ensuring that educational resources and initiatives reach all segments of society and addressing disparities in access to information and resources.

Behavioural change, education, and community engagement are integral in fostering a collective understanding of environmental challenges and promoting sustainable actions. Encouraging individuals and communities to adopt eco-friendly practices and promoting climate literacy are key strategies in reducing emissions and building a more sustainable society.

Adaptation and resilience are crucial in addressing the impacts of climate change. Here's an overview of adaptation strategies, building climate-resilient communities, and ecosystem-based adaptation techniques:

1. Adaptation Strategies to Climate Change

Infrastructure and Planning:

Climate-Resilient Infrastructure: Designing and building infrastructure that can withstand extreme weather events and rising sea levels.

Urban Planning: Implementing zoning laws and land-use planning that consider climate risks to minimize vulnerability in cities.

Water Management:

Flood Protection Measures: Constructing levees, dams, and flood barriers to mitigate flood risks in vulnerable areas.

Water Conservation: Implementing water-saving technologies and strategies to address water scarcity during droughts.

Agriculture and Food Security:

Crop Diversification: Promoting diverse crop cultivation to adapt to changing climatic conditions and reduce vulnerability to crop failures.

Climate-Resilient Farming Practices: Implementing soil conservation, water management, and agroforestry techniques to enhance resilience in agriculture.

2. Building Climate-Resilient Communities

Community-Based Approaches:

Early Warning Systems: Developing and implementing systems to alert communities about impending extreme weather events or natural disasters.

Capacity Building: Training communities to respond effectively to climate-related risks and emergencies.

Infrastructure and Shelter:

Resilient Housing: Constructing or retrofitting homes to withstand extreme weather events like storms, hurricanes, or wildfires.

Community Infrastructure: Strengthening community facilities like schools, hospitals, and emergency shelters to serve as resilient hubs during disasters.

3. Ecosystem-Based Adaptation Techniques

Nature-Based Solutions:

Ecosystem Restoration: Rehabilitating degraded ecosystems, such as wetlands, mangroves, and forests, to enhance natural resilience.

Green Infrastructure: Using natural elements like vegetation and permeable surfaces in urban areas to manage stormwater and reduce flooding.

Biodiversity Conservation:

Preserving Biodiversity: Protecting diverse ecosystems and species to maintain ecological balance and resilience against climate change impacts.

Integrated Ecosystem Management: Employing management practices that consider entire ecosystems and their interactions.

Importance and Challenges

Importance of Adaptation:

Risk Reduction: Building resilience helps reduce vulnerability to climate-related risks and minimizes damages from extreme events.

Sustainable Development: Adaptation measures support sustainable development by safeguarding communities, ecosystems, and economies.

Challenges:

Resource Constraints: Limited resources and funding pose challenges in implementing large-scale adaptation projects, especially in vulnerable regions.

Complexity and Interconnectedness: Addressing multifaceted climate risks requires integrated approaches and collaboration across sectors and stakeholders.

Adaptation and building resilience are essential components of climate change response. Implementing diverse adaptation strategies, fostering community resilience, and leveraging ecosystem-based approaches are integral in reducing vulnerabilities and enhancing the capacity of communities and ecosystems to cope with the impacts of a changing climate.

Did you love *Emerging Trends in Carbon Emission Reduction*? Then you should read *Book of Herbs and Spices* by J K Arora!

"Book of Herbs and Spices" is a captivating book that uncovers the remarkable healing properties hidden within nature's kitchen. From the anti-inflammatory prowess of turmeric to the calming effects of chamomile, this insightful guide explores the medicinal potential of herbs and spices. It goes into their historical use in traditional medicine and offers practical advice on harnessing their therapeutic benefits in modern health and wellness. This book is a valuable resource for anyone interested in natural remedies, offering a wealth of knowledge on how these humble ingredients can contribute to overall well-being and a healthier lifestyle. The book has over 80 herbs and spices commonly used in all countries of the world. Herbs and spices are not property of any one country and have been used by several ancient civilizations all over the world and their origin depends on the climatic conditions of those countries and not on any particular civilization.